Greening through IT

Greening through IT

Information Technology for Environmental Sustainability

Bill Tomlinson

The MIT Press
Cambridge, Massachusetts
London, England

For information about special quantity discounts, please email <special_sales@mitpress.mit.
edu>.

This book was set in Stone Sans and Stone Serif by the MIT Press.
Printed on recycled paper and bound in the United States of America.

Library of Congress Cataloging-in-Publication Data

Tomlinson, Bill, 1972–
Greening through IT : information technology for environmental sustainability / Bill Tomlinson.
    p. cm.
Includes bibliographical references and index.
ISBN 978-0-262-01393-2 (hardcover : alk. paper)
1. Green technology. 2. Information technology—Environmental aspects. I. Title.
TD145.T48   2010
303.48'3—dc22

                                                                                    2009033821

    10   9   8   7   6   5   4   3   2   1

For Miles, who kicks.

Contents

Acknowledgments

Bringing a book to completion involves contributions and energy from dozens of people. Each of these people mentioned here played a unique role in the process of developing this text. I am grateful for their help.

In particular, Doug Sery, Margy Avery, Marcy Ross, Katie Helke, and everyone else at the MIT Press were encouraging and supportive through all stages of the process. Paul Dourish, Bonnie Nardi, Six (Michael) Silberman, Michael Patrick Johnson, Michael Riccobono, Eric Kabisch, and several anonymous reviewers provided valuable feedback on drafts of the manuscript. Discussions and interactions with Don Patterson, Gillian Hayes, Susan Sim, J. Michael Moshell, Larry Smarr, F. Lynn Carpenter, Vibeke Sorensen, Eli Blevis, Elaine Huang, Jen Mankoff, Lisa Nathan, Batya Friedman, Phoebe Sengers, Jay Hasbrouck, Celia Pearce, Andrew Torrance, Ken Russell, David Filippi, Bill Bainbridge, Richard Anderson, Jon Kolko, Katie Clinton, Linda Ward, and Steve Price helped improve the ideas presented in the book.

Many students in my research group, the Social Code Group, worked on projects described in chapters 6–8 and contributed to conversations about the book's content in our meetings. In particular, I would like to thank Eric Baumer, Man Lok (Simon) Yau, Joel Ross, Nadine Amsel, Faith Dang, Sara Goetz, Bruno Nadeau, Ajey Shah, So Yamaoka, Lorenzo Canales, Andrew Correa, Grace Chiang, Kenneth Chua, Ellen Eramya, Bryant Hornick, Zack (Gang) Ji, Paul Mac Alpine, Stefan Marinov, Uel (Jack) McMahan III, Robert Moodey, Erika Ramos, Michael Riccobono, Martin Schmidt, Nitin Shantharam, Anju Sharma, Robert Simpson, Jordan Sinclair, Craig Yoho, and Andrew Zaldivar. Undergraduates Mark Sueyoshi and Andrew Zaldivar provided research assistance for several sections of the book. Rob Olson, Jeff Greenberg, and Lilly Irani improved the Trackulous project. Riley Pratt, Kristin Young, and Calen May-Tobin helped our group learn more about ecology throughout our collaboration on the EcoRaft project. The students in the Informatics 235 course I taught in spring 2008 as well as the several Informatics 161 courses in 2007–2008 contributed to refining many of the concepts in the following pages. Several chapters build on projects and coauthored papers created by collaborators in the above lists. Specific contributors are credited at the beginning of the section inspired by their work.

The ability to write a book of this kind is made vastly easier by an effective intellectual context and support structure. Debra Richardson, David Redmiles, Marty Beach, Sabina Chung, Barbara Marr, and the University of California, Irvine's Informatics Department faculty, staff, and students as well as Larry Smarr, G. P. Li, Albert Yee, Stu Ross, Shellie Nazarenus, Andrea Bell-Wheelans, Anna Lynn Spitzer, and the other faculty, staff, and students at the California Institute for Telecommunications and Information Technology (Calit2) provided an intellectual and logistic home for this work.

The research and writing of this book was supported by a variety of organizations and sources, including the Donald Bren School of Information and Computer Sciences, Calit2, the Alfred P. Sloan Foundation, the Nicholas Foundation, Microsoft, Emulex, and the University of California, Irvine, through a Faculty Career Development Award. This material is based upon work supported by the National Science Foundation under Grant No. 0644415. Any opinions, findings, and conclusions or recommendations expressed in this material are those of the author and do not necessarily reflect the views of the National Science Foundation.

My family has always encouraged academic pursuits, especially those that deal with environmental themes. My parents, Mickey and Madaline Tomlinson; my sister, Lynn Tomlinson; her husband, Craig Saper; their kids, Sam and Lucy; my uncle, Bill Borrebach, and aunt Landra Prins, have discussed the book with me and helped make it possible in many different ways. Thank you all.

Finally, I wish to express my profound gratitude to my wife, Rebecca Black. She supplied me with a firsthand demonstration of how to write a book (Black, 2008), talked with me about these ideas during many walks along the Back Bay of Newport Harbor, read countless drafts, and was an endless source of enthusiasm and encouragement. Without her love and support, this book would not have been possible.

1 Introduction to Green IT

Kerala, a state in the southwestern part of India, has a substantial fishing industry, directly employing more than a million people. Sardines are one of the main species that the commercial fishers catch. Starting in the late 1990s, mobile phone service became available in Kerala. The fishers gradually began using the phones to find customers in need of their catch and coordinate sales. In previous years, 5 to 8 percent of the daily catch of sardines was wasted, with sellers sometimes unable to find buyers. With the introduction of mobile phones, this waste was eliminated completely, as the fishers and the buyers dynamically streamlined their markets. The fishers' profits increased, consumer prices declined, and fewer sardines went to waste unnecessarily (Jensen, 2007).

Utility companies and research laboratories around the world are working to develop and install smart electricity and gas meters in millions of households and businesses. These meters are designed to support real-time energy pricing and enable households and businesses to optimize their usage. For example, various energy-intensive activities such as running a clothes dryer could be scheduled at times when electricity demand is low. These meters will simultaneously save money for consumers and allow the utility providers to deliver energy more efficiently.

On July 14, 2007, a group of fifty-seven naked bicyclists took to the streets of Seattle, Washington. This event was one of many such rides in cities across the globe as part of the "World Naked Bike Ride." The organization's Web page offers the following credo: "We face automobile traffic with our naked bodies as the best way of defending our dignity and exposing the unique dangers faced by cyclists and pedestrians as well as the negative consequences we all face due to dependence on oil, and other forms of non-renewable energy" (WorldNakedBikeRide.com, 2007). Rather than relying on paper brochures or other traditional media, the organization used the World Wide Web, email, and other information technologies to support the planning, coordination, and deployment of events around the world. In the online wiki, a comment for the Seattle event reads: "Great weather, 3 arrested" (NakedWiki.org, 2008).

The above examples illustrate just a few of the many ways in which environmental issues and information technology (IT) are beginning to interact.[1] From Cisco's use of telepresence to reduce corporate travel (Hsieh, 2009) to the large-scale migration of computing processes to places where energy is available (Hopper & Rice, 2008), instances of IT systems having an environmental impact are emerging across a wide range of domains. This book seeks to provide a framework for thinking about IT systems that address environmental topics. By doing so, it aims to help people work more effectively, both individually and together, to address the numerous environmental concerns currently facing the planet.

The Field of Green IT

Over the last several years, the term "Green IT" has begun to be used to describe a field at the juncture of two trends. The first trend involves the growing concern about environmental issues across many human communities. For example, a *Newsweek* poll in August 2007 found that 61 percent of adults in the United States said they had personally taken "steps to reduce [their] own energy consumption because of concerns related to climate change and global warming" (Polling Report, 2008). A survey of 49,243 teens from around the world revealed that 74 percent of them "believe that global warming is a serious problem and are more concerned about it than any other issue including drugs, violence or war" (Greenpeace, 2007). As further evidence, the film *An Inconvenient Truth* won the 2007 Academy Award for best documentary feature, and Al Gore shared the 2007 Nobel Peace Prize with the UN Intergovernmental Panel on Climate Change "for their efforts to build up and disseminate greater knowledge about man-made climate change, and to lay the foundations for the measures that are needed to counteract such change" (Nobel Foundation, 2008). These and many other recent cultural phenomena reflect a growing awareness of the impact that humans and our activities have on the world. The term "Green" has become connected with this sense of awareness; the word choice is based on the primary color of plant leaves and reflects a concern for the ecosystems in which humans exist.

The second trend involves IT—the use of digital tools and techniques for manipulating information, and the social phenomena that surround these systems. (Actually, two acronyms, IT and ICT, could apply here. As noted above, IT stands for information technology and is broadly known across a range of public audiences. ICT stands for information and communications technology; this more explicitly inclusive term is used in certain fields of academia and a range of government/economic contexts. To make the content of this book broadly accessible, the acronym IT is used throughout.)

IT is growing at a rapid rate; as an example, while mobile phones were relatively rare a decade ago, now more than half the world's people have them (Reuters, 2007). There are many different forms of IT, in terms of both the information systems that exist—such as software packages, databases, and networks—and the devices that people and institutions use to access these systems—such as notebook computers, servers,

mobile phones, and personal digital assistants. Humans inhabit the growing information ecologies that result from this abundance of devices and systems (Nardi & O'Day,
1999). The rapid adoption of IT is transforming societies around the world, affecting
many different aspects of human life from communication between individuals to the
workings of international politics.

Green IT brings together these two areas, environmental issues and IT, and explores
the ways in which they connect to each other. In particular, it examines the opportunities for IT to address issues related to the global ecosystem. The rapid growth and
acceptance of IT worldwide suggests that this area may be a fruitful one in which to
seek possibilities for environmental change, building on other transformative societal
effects already underway. Andy Hopper (2008), professor of computer technology and
head of the computer laboratory at the University of Cambridge, states the case suc-
cinctly: "Computing is a crucial weapon in our armoury for ensuring the future of the
planet" (p. 1).

It is important to note that not all facets of IT are environmentally favorable.
Computation is beginning to occupy a nontrivial amount of the power consumption
around the world, and electronic waste (e-waste) is a fast-growing concern. In addition,
the culture of disposability implied by the rapid innovation and turnover rate in computational technologies is deeply problematic. According to a report by the Gartner
research firm, the IT sector emits CO_2 at a rate approximately equivalent to that of the
airline industry (Mingay, 2007). This rate is growing rapidly, especially in the mobile
computing area. Nevertheless, according to *Smart 2020*, a report by the Climate Group
(2008) on behalf of the Global e-Sustainability Initiative, the potential positive environmental benefits enabled by IT are five times as great as the environmental footprint
of IT itself. Given the inherent complexity of environmental issues and the inefficient
ways humans currently live, innovations in IT across many aspects of society can have
a strongly positive net environmental impact.

The environmental opportunities presented by IT systems are both external and
internal to the IT field itself. Externally, Green IT can have an effect on areas from economics (von Weizsäcker, Lovins, & Lovins, 1998) to ecological monitoring (Szewczyk
et al., 2004) to the details of people's everyday lifestyles (Mankoff, Matthews, Fussell,
& Johnson, 2007; Woodruff, Hasbrouck, & Augustin, 2008). Internally, Green IT can
help reduce the impact of e-waste (Grossman, 2006; Hightower, Smith, Sonnenfeld, &
Pellow, 2006), enable sustainable interaction design (Blevis, 2007), and decrease energy
consumption by computational systems (Markoff, 2006). While the term "Green IT" is
sometimes used to refer solely to the ways in which the IT industry itself can become
more sustainable, in this book Green IT involves any IT that helps to make any aspect
of society greener, including IT itself. The title of this book, *Greening through IT*, seeks

to capture this focus on the use of IT across many different sectors, rather than concentrating just on its own impact.

The ultimate goal of this text is to help humans and the ecosystems in which we live achieve a state of sustainability. Various definitions have been offered for the term "sustainability"; for example, Ismail Serageldin (2006) offered a human-centric definition, stating that "sustainability is giving future generations as many opportunities as, if not more than, we have had ourselves." For the purposes of this book, sustainability is a characteristic of a system—in this case, the global ecosystem—in which all defining processes, such as the maintenance of biodiversity (including *Homo sapiens*) at a high quality of life, are able to continue indefinitely. A sustainable system can be juxtaposed with one in which limitations on the availability of resources, overproduction of waste, or other factors will eventually cause the system to collapse.

Other useful terms have been put forward to describe related concepts. "Resilience" has been defined as "the capacity of a system to absorb disturbance and reorganize while undergoing change so as to still retain essentially the same function, structure, identity, and feedbacks" (Walker, Holling, Carpenter, & Kinzig, 2004, p. 1). Aldo Leopold (1966) offers a more general term, "right," suggesting: "A thing is right when it tends to preserve the integrity, stability, and beauty of the biotic community. It is wrong when it tends otherwise" (p. 262). Because of its existing association with an array of ongoing international efforts, the term "sustainability" will be used to represent this suite of concepts throughout this text.

Abundant scientific evidence—for example, the findings in the IPCC's (2007b) Climate Change 2007 report—suggests that humanity is not currently living in a sustainable way. Fortunately, people around the world are increasingly becoming aware of environmental issues and taking action to address them. Despite this growing engagement, we need to ramp up the intensity of these efforts in order to avoid profound and potentially catastrophic global change. The coming chapters explore the critical role that IT can play in this undertaking. For the past several millennia, technology has multiplied the rate at which humans brought about the current environmental concerns; by broadening humans' horizons of time, space, and complexity, Green IT can provide a multiplier effect on our ability to resolve those same environmental issues in the future.

Environmental Issues Operate on Broad Scales of Time, Space, and Complexity

Humans and all other organisms require certain characteristics of our habitats to remain within certain parameters in order for us to survive. If our body temperature goes too far from 98.6°F, if we do not receive air, food, or water for a certain period of time, or if many other conditions are not met, we cannot continue to exist. Throughout this text, the term "environmental" will refer to the collection of topics that relate to the nonhuman physical and biological context in which humans live, and on which we rely for our continued existence.

The key problem in the way humans understand and act on environmental issues is one of scale. Environmental issues tend to occur on broad scales of time, space, and complexity compared to the typical scope of human concerns. These differences are particularly apparent in domains such as population, resource use, waste production, and species extinction. Regarding the first of these areas, the world population is approximately 6.7 billion people and has been increasing rapidly, doubling within the last forty years. People inhabit over 190 countries, speak thousands of languages, and live in cities with millions of inhabitants. Within this century, the world population is likely to grow to over 9 billion (United Nations, 2004).

People are exhausting natural resources, many of which regenerate slowly through nonhuman processes, on a global scale; nevertheless, we do not adequately address the restoration of these resources. For example, humans consume oil and other petroleum products at a rapid rate, but these resources only come into existence naturally via geologically long periods of compression and heating. "Peak oil" is the term given to the point at which worldwide petroleum production will reach its maximum level; this point is seen as occurring within the next few decades, if it has not already happened (U.S. Government Accountability Office, 2007). Peak oil indicates the dramatic imbalance between the time scales of petroleum formation and exploitation.

As another example of humanity's vigorous resource use, consider that half the world's forests, which once covered 48 percent of Earth's surface, have disappeared during human history (NRDC.org, 2004). The loss of these forests has a variety of effects, such as reducing the planet's ability to convert CO_2 to breathable O_2 and eliminating the habitats of many biological species. Beyond the impacts we can already see, the long-term environmental effects of human civilization's current actions will take decades or centuries to emerge fully. Therefore, it is likely that increasing environmental degradation as a result of human resource consumption will be with us for a long time.

In the process of using these resources, humans produce large amounts of waste as well. For example, in 2005, the world produced over 28 billion metric tons of CO_2 from the consumption of fossil fuels (U.S. Environmental Protection Agency, 2007); CO_2 is one of the greenhouse gases (GHG) largely responsible for global climatic disruption. (John Holdren [2008], Barack Obama's science and technology adviser, prefers the term "global climatic disruption" to "global warming.") According to the Intergovernmental Panel on Climate Change (IPCC, 2007b), "Most of the observed increase in global average temperatures since the mid-20th century is very likely due to the observed increase in anthropogenic greenhouse gas concentrations" (p. 10).

Global climatic disruption is the most famous problematic result of this waste production, but there are many others. According to the U.S. Environmental Protection Agency (2008a), the United States produces 230 billion kilograms of municipal solid waste per year, or more than 2 kilograms per person per day. In 2006, people around the world consumed 47 billion gallons of bottled water; the United States had the highest per capita consumption of bottled water, at more than 29 gallons per person that year (BottledWater.org, 2006). Only 31 percent of these bottles were recycled; the others ended up in landfills or other locations (U.S. Environmental Protection Agency, 2007). Taken together, these unsustainable habits represent a major issue that needs to be addressed. Nevertheless, the impacts of these choices are rarely immediately apparent, but instead unfold over long periods of time, at great distances away, and in a variety of indirect ways.

The drastic increase in worldwide species extinctions over the last several hundred years also suggests that we are not on a sustainable trajectory. According to a report by the Secretariat of the Convention on Biological Diversity (2006), a treaty signed by 150 world government leaders at the 1992 Rio Earth Summit, "Species are going extinct at rates 1,000 times the background rates typical of Earth's past." This trend is sometimes referred to as the "Sixth Extinction" (Leakey & Lewin, 1996)—the sixth major extinction event since the beginning of life on Earth, and the most severe since the extinction of the dinosaurs sixty-five million years ago. The ecological factors that underlie each species' extinction are complex; taken together, the complexity involved in the extinction of thirty to fifty thousand species per year (Leakey & Lewin, 1996; Olson, 2005) is difficult to grasp.

The extinction of species is significant not just for whatever inherent value one might place on the species themselves. Additionally, the sustainable survival of humans is more likely if there are many different kinds of life in existence. Diversity offers robustness in the face of perturbation. Diverse ecosystems are able to absorb and adapt to change more easily than homogeneous ecosystems. There are numerous examples of the fragility of monocultures (i.e., large tracts of the same crop), such as the potato blight that caused the Great Irish Famine in the mid-nineteenth century, and the spread of the boll weevil across cotton farms in the southeastern United States in the early twentieth century. Humans are part of the ecological complexity of Earth. While the world may be made temporarily more amenable to humans by the dedication of large areas of land to corn, palms, and cows, our tendency to exploit a few species on large scales leads to the extinction of many others, and thereby threatens the stability of the global ecosystem as a whole.

The environmental issues currently facing the planet are substantial. Nevertheless, people, governments, and other human institutions often choose to attend to other, more immediate, local, or accessible concerns, rather than addressing any of the above problems with vigor.

Humans Operate on Narrow Scales of Time, Space, and Complexity

In the above brief account of some of the world's environmental issues, I used a common shorthand in discussions of scale: whenever I used an adjective like "large," "slow," or "big," I assumed that the reader would understand that I meant each of these terms "from a human perspective." The amount of CO_2 humans produce is not large when compared to the mass of the Sun. The process of forest regeneration is not slow when compared to continental drift. The extinction of a single species is not complex compared to the evolution of all life. Nevertheless, to a person trying to make sense of these issues, the scales involved are larger than those we typically confront. Environmental issues tend to occur on much broader scales than humans are skilled at comprehending.

Human anatomy, physiology, and behavior evolved to deal with a certain set of problems that recurred frequently in our ancestors' lives. They looked for food and water; they avoided danger; they sought opportunities for reproduction; and they cared for their children. All of these interactions usually happened on what I will refer to as "human scales" of time, space, and complexity. Individual humans rarely worried about gathering food a thousand years after their deaths. They did not concern themselves with a boulder bounding down a mountain in a different continent; they worried about the ones over their heads. And in helping their children learn and grow, they taught them about the world they would encounter every day, so that they too might live well.

As humans developed more and more complex societies, moving to unprecedented forms of communication, memory, and collective action, cultural factors began to broaden their horizons. People developed rules and laws to enable them to plan further into the future and live together at larger scales. They told stories about their past encounters so their social partners could learn from their experience without needing to suffer any negative outcomes themselves. They developed governments and corporations so that they could together reap the benefits of broader-scale activity. Economic innovations such as discount rates enabled them to formalize the way value changes over time. City walls and political borders helped humans understand the limits of their physical domains. Organizational hierarchies and division of labor let people manage the growing complexity of their lives.

Human scales of time, space, and complexity are broad, from some perspectives. Our lives are long when compared to those of ants; our bodies are big when compared to bacteria; and our ability to comprehend is vast when compared to that of a macaque. Nevertheless, the topic of this book is environmental change, and the relative narrowness of human comprehension and action when compared to the scale of the world's environmental concerns is a critical factor in our ability to respond effectively. "Business as usual"—that is, continuing along current trajectories of population, resource consumption, waste production, and global impact—with our current horizons of time, space, and complexity, is not sustainable.

Human institutions such as corporations and governments have many of the same characteristics that people have, but on somewhat different scales of time, space, and complexity. In fact, many are explicitly organized to help leverage the benefits of broader scales of operation. For example, Ronald Coase (1937) won the 1991 Nobel Prize for Economics based on his paper "The Nature of the Firm," which shows how reduced transaction costs enable firms to outperform individuals operating as independent contractors in a free market; this reduction in complexity helps a firm succeed in business more effectively than a collection of individuals. Similarly, governments help smooth out problems that might be caused by individuals acting out of pure self-interest without regulation. Governments thus allow people to operate with longer time horizons and move across larger areas of space with an expectation of consistent interactions with each other.

At the same time, corporations, governments, and other institutions still primarily serve human-scale interests. Corporations are tasked with enhancing shareholder value. The U.S. Constitution begins: "We the People of the United States, in Order to form a more perfect Union, establish Justice, insure domestic Tranquility, provide for the common defence, promote the general Welfare, and secure the Blessings of Liberty to ourselves and our Posterity, do ordain and establish this Constitution for the United States of America" (archives.gov, 2009). This document is ultimately a human-centered text, and as such reflects human scales of time, space, and complexity. Human institutions, seeking to leverage opportunities to act with broader horizons to serve human needs, often lie somewhere between the scale of individual humans and that of environmental concerns.

IT Compresses Time, Space, and Complexity

Technology, our ability to work with tools and develop techniques for manipulating our environment to better serve our liking, developed alongside human communities and has a long history of extending our reach. We invented agricultural implements, rapid transportation systems, and powerful weapons. Technology has provided a "force multiplier" in many domains of human experience, letting us act with greater impact for less effort. Through the technologies we have developed, humans have made our own horizons broader, crossing oceans, forming multinational organizations, and watching light arrive from long-dead stars.

IT in particular has proved to be effective at broadening our horizons, or seen another way, compressing time, space, and complexity. IT compresses time in many ways— for example, by storing abundant information for later retrieval, letting us model the past and predict the future, and enabling the synchronization of many different human activities. IT compresses space by allowing us to communicate over great distances, browse maps of the entire world, and transport goods and people around the globe. IT compresses complexity by augmenting our memories, allowing devices to perform repetitive calculations, and establishing agreed-on standards for the cooperation of devices and people. Across human history, IT systems have been critical in broadening the scales on which humans operate. We need systems that broaden our understanding and ability to take action across broad ranges of time, space, and complexity; framed another way, we need compression algorithms that shrink global environmental problems to human-comprehensible scales.

In terms of human institutions rather than just individuals, IT helps bridge gaps in scale as well. IT has been used to provide more comprehensible information to individuals about a variety of institution-scale phenomena, from stock prices to electricity rates to election results. IT also allows institutions to act more effectively based on expert opinions and analyses predicated on computational simulation, modeling, and visualization. IT has the potential to extend both of these connections, bridging from human scales to institutional ones, and institutional scales to environmental ones, thereby indirectly enabling individuals to impact environmental concerns via the mediation of institutions.

Green IT Bridges from Human Scales to Environmental Scales

Technology and IT have not always worked to the benefit of environmental sustainability. In fact, quite the contrary: technological innovations have allowed us to farm, burn, mine, fish, drive, reproduce, and buy to the detriment of many environmental concerns. From these problems, though, arise numerous opportunities for people to make the way we live more sustainable. Helping people and institutions discover, understand, and act on these and other environmental possibilities is the primary goal of Green IT.

By helping spread information about environmental issues, and enabling humans and our organizations to share best practices for addressing them, Green IT can hopefully assist in mitigating the global problems that appear to be looming in our future. Many examples of explicitly environmental IT systems already exist—from smart energy grids to systems that optimize hybrid car engines (Kleimaier & Schröder, 2004) to the Earthday Network's Ecological Footprint Quiz (Earthday.net, 2008). In addition there are many IT systems that have been developed for nonenvironmental reasons, but that have implicit environmental impacts, such as GPS systems and online mapping software, which lead to more efficient travel and therefore reduced CO_2 emissions. This book draws on existing examples of systems whenever possible, in order to ground the discussion in real-world applications. Green IT also shows promise as an area in which future developments will enable new ways of living and working together. In this spirit, the book presents ideas for novel innovations in the field of Green IT.

The ways in which IT can benefit environmental issues take a variety of forms. One axis along which these innovations may be arrayed is from "personal" to "institutional." Personal IT systems enable individuals and small groups to broaden the horizons of time, space, and complexity with which they think and act, thereby enabling them to respond more effectively to a range of environmental concerns (see figure 1.1a). Institutional IT systems have a similar effect, but broaden the horizons of understanding and action for corporations, governments, nongovernmental organizations (NGOs), universities, and other large-scale organizations (see figure 1.1b). These two forms of Green IT are mutually reinforcing.

Personal

Green IT can help individuals participate in many different ways to address the world's current environmental concerns. For instance, it can provide information that encourages people to exert more effort in this direction. This form of participation can be simple—for example, reading a post on an environmental Web site about walking up the stairs instead of taking the elevator—or more dramatic—such as selling one's car, and then using a bicycle or a car-sharing system such as Zipcar instead.

IT can also improve the efficiency with which we enact our daily lives. Using a Garmin Nuvi GPS navigation device to take a more optimal route among several chores

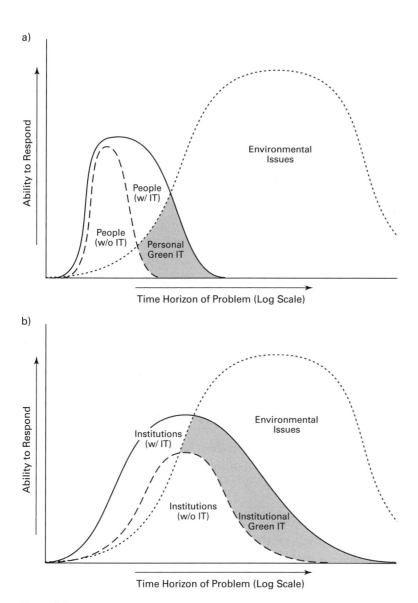

Figure 1.1
IT extends the ability of people (a) and institutions (b) to respond to environmental issues.

in an afternoon or hosting a business meeting via teleconference rather than flying people in from other countries are examples of more efficient practices that are more environmentally conscious as well. While there may be accompanying problems with these and other innovations, such as the loss of serendipity in driving and reduced benefits of face-to-face interaction, we are reaching a point where environmental concerns should weigh heavily in the choices we make.

Furthermore, Green IT can provide opportunities to address environmental issues through improved collaboration. For example, CouchSurfing.com allows people to plan trips where they stay in someone's house rather than a hotel. In addition to reducing the environmental impact for the trip, it also helps enable a social reward through the connections and possible friendships established among like-minded individuals. This and many other online systems help groups of people organize themselves and find strength in numbers for environmental goals. Earlier in this chapter, I mentioned a poll suggesting that 61 percent of U.S. adults have taken steps to reduce their energy consumption due to global warming. Put another way, 39 percent of U.S. adults have not taken any steps at all to reduce their energy consumption. If sustainability is the goal, there is still a long way to go. Understanding the mechanisms by which Green IT can lower the bar for participation and increase the rewards that come from greater effort could be an important contribution to a sustainable future.

Institutional

Beyond the actions of individuals, there are abundant ways that Green IT can contribute through corporate, governmental, and other institutional efforts. Green IT innovations can extend the scale with which human institutions support the movement toward global sustainability.

One of the primary institutional contributions of Green IT is through improved infrastructures. Smart energy grids enable more efficient power utilization. Improved transportation systems reduce fuel use while optimizing the movement of people, objects, and materials around the world. More effective waste management systems can facilitate more comprehensive recycling and salvaging of useful materials after their initial usage is complete.

Green IT can also improve large-scale planning and policy decisions. IT is helping scientists and other experts develop an understanding of many real-world phenomena—a critical factor in making informed, far-reaching decisions. Visualization and modeling techniques can help policymakers understand scientific factors and likely outcomes more fully, thereby allowing for more informed decisions. E-government systems can involve the populace more effectively in policymaking, providing citizens with information about why decisions are made, and encouraging them to participate in community- or worldwide efforts.

Institutional action through Green IT also has the potential to enable individuals to live more sustainably on a large scale. It is often difficult to make environmentally sensitive choices when those choices are discouraged by law or lack of infrastructure. Understanding the ways in which human institutions can provide individuals with environmental options is crucial to the goal of achieving global sustainability.

It is important to note that there is fluidity between personal and institutional Green IT innovations. For example, the smart energy meters mentioned at the beginning of this chapter allow individuals to monitor their consumption habits and institutions to analyze usage across millions of households, both of which help humanity live more sustainably. While there is not a clear distinction, personal and institutional forms of Green IT address different scales of time, space, and complexity. These efforts are complementary; if some of the personal Green IT innovations allow people to take environmental actions in domains previously under the aegis of institutions, it now frees up those institutions, which operate under constraints of available time and energy, to focus on broader environmental issues.

An Extended Human-Centered Approach

This book is grounded in the field of human-centered computing (HCC), and takes a human-centered approach to the Green IT field. HCC focuses on humans as the most important element in the development of computing systems. Moving beyond narrowly scoped HCC, however, *Greening through IT* presents an extended view of human-centeredness, looking at how computing can situate human civilizations at the locus of responsibility for environmental issues, and enable individuals and groups to act to resolve these issues. This human-centered approach is distinct from anthropocentrism and is not meant to downplay the significance of the survival of other species; rather, it sees humans as the species with the greatest potential to orchestrate a coherent response to the world's current environmental problems, and therefore the most expedient path to addressing those issues.

HCC arose from a response by computing professionals and researchers to an earlier state of computing. During its first few decades, computing resources were scarce, and hence humans were expected to take a great deal of effort to organize and customize content intended for computers prior to its being input into the machine. From large stacks of punch cards to complex command-line systems, interfaces to computers were complex and unintuitive. In the last several decades, though, as computing power has become more abundant and less limiting, there has been a move toward increased human usability in computing. Humans began to have more active roles in human-computer systems (Bannon, 1992), and to be included in the process of designing interfaces. As computers have reached a mass market, designing interfaces for ease of use rather than just computational efficiency has become increasingly important to the computing industry.

HCC continues to be a growing field, with programs at the U.S. National Science Foundation (NSF), the U.S. National Aeronautics and Space Administration, and several major research universities. NSF offers the following definition of the area:

[HCC] research encompasses a rich panoply of diverse themes in Computer Science and IT, all of which are united by the common thread that human beings, whether as individuals, teams, organizations or societies, assume participatory and integral roles throughout all stages of IT development and use. People design new technologies; people, in teams and organizations, at school and at home, use them; people anticipate and enjoy their benefits; and they learn about the outcomes of use (whether anticipated or not) and translate that knowledge into the next generation of systems. At the same time, new information technologies and human societies co-evolve, transforming each other in the process. As a consequence, the design of IT must be sensitive to human values and preferences. (U.S. National Science Foundation, 2006b)

A human-centered approach leads to several benefits. It produces computing systems that allow people to work more effectively and play more enjoyably. It develops interfaces that mesh smoothly with people's abilities and desires. It allows for communication and collaboration on previously unheard-of levels. Through these and other advantages, a human-centered approach contributes to greater human happiness.

Extending the focus of HCC to consider factors on a broader set of horizons than the usual scope of individual humans provides the central structure around which this book is organized. This "extended human-centered computing" (EHCC) perspective provides a way of understanding the potential for Green IT. This view discards the notion that the immediate gratification of human desires is an unconquerable problem for the world's ecosystems.

A broader set of horizons for HCC is necessary if the world's environmental concerns are to be addressed effectively. There is already some awareness in the research community of this need. For example, NSF includes "broader impacts" as one of its key evaluation criteria for grant making. Yang Li and James Landay (2008) offer that their theoretical framework "uses a larger unit (long-term, high-level activities such as staying fit) for analysis and design than simple tasks (such as using a treadmill), which have been the focus of traditional approaches." Similarly, as mentioned previously, Eli Blevis's (2007) focus on sustainable interaction design seeks to give computing systems greater longevity in a variety of ways. The "value-sensitive design" approach put forward by Batya Friedman and her colleagues posits a need to "broaden the goals and criteria for judging the quality of information systems to include those that advance human values" (Friedman & Freier, 2005, p. 371). Value scenarios (Nathan, Klasnja, & Friedman, 2007) can help support long-term thinking about important design problems. Many other research projects focusing on broad-horizon environmental issues will be discussed throughout this text. Focusing on the impact of IT systems a century or millennium from now, or on a worldwide scale, may seem impractical, given people's predilections for thinking of our own immediate, local issues first. Certain ways of thinking about computing, however, may help shift the focus in order to serve the same goals as a human-centered perspective with a broad horizon in time, space, or complexity.

The human-centered approach taken here also embraces human institutions such as corporations, universities, governments, and other organizations. These institutions were formed by humans and are staffed by humans; as such, they often reflect a human-centered approach. They tend to have broader horizons than individuals; however, they tend to focus on problems facing groups of people. In order for most institutions to deal effectively with environmental concerns, they too need to broaden their horizons still further in time, space, and complexity. While my research over the past several years has tended to focus on personal Green IT, and therefore many of the examples throughout this book are from that domain, the core principle of this text—that IT can help make connections between human-scale action and broader environmental concerns—applies to human institutions as well as individuals.

Broadening Time

While NSF's vision for HCC does include some attention to human values such as environmental sustainability and social justice, and thus has some connection to broad-scale undertakings, there is a pervasive rhetoric associated with HCC that leads to short-term satisfaction of human users' needs and desires. I use "short-term" here in the economic sense, as in short-term capital gains, rather than in the cognitive science sense, as in short-term memory. Unless otherwise specified, I will use "short-term" to mean hours, days, weeks, or months, and "long-term" to mean years, decades, centuries, and millennia.

Relatively short-term human satisfaction has been a primary focus for much HCC-related scholarship. For example, citing Mark Weiser's seminal article, "The Computer for the 21st Century," Li and Landay (2008) summarize a prevailing view within the research community as follows: "Ubiquitous computing (ubicomp) promises to support our everyday activities by weaving computing power into the fabric of everyday life." (p. 1303). Even the term "human-centered computing" (rather than "humanity-centered computing," "world-centered computing," or many other possible choices) suggests a focus on the interests of an individual person.

A concentration on narrowly scoped human centeredness makes sense when the development of computing is seen in the context of human cognitive evolution. Humans are exceedingly skilled at thinking about scales of time that reflect our own needs for survival and reproduction. On the evolutionary landscape in which we evolved, our ancestors could neither gather information effectively on global or millennial scales, nor act effectively on them. Except for our ability to use technology, which is good for all sorts of things, human brains are not particularly well suited for comprehending or acting on broad-scale phenomena. Technology, though, has an established tradition of helping us deal with scales of time beyond those in our comfort zone. IT in particular, with its ability to collect, analyze, and store vast bodies of data with a high degree of precision, support communication and collaboration across time and space, and interact with people and control devices, is appropriate for supporting human efforts to understand broad issues and coordinate effective responses.

A focus on short-term human usability and support in computing also makes sense given the computing field's history of interfaces that required humans to alter their behavior profoundly to accommodate machines' needs. Moreover, there is nothing inherently wrong with attending to the short-term needs and desires of users. In fact, if their short-term needs and desires are not attended to, users will frequently abandon a system. However, a key issue here is that it is insufficient to concentrate *only* on short-term issues. Computing researchers and professionals need to look at long-term issues as well.

Making it more socially acceptable to think about issues on a longer time scale could help. People are capable of delayed gratification—that is, making decisions that are costly in the short term but beneficial in the long term. Savings accounts, dieting, and university educations could all be seen as examples of this phenomenon. Because of differences in how people perceive the value of present benefits versus the value of benefits in the future, individuals may disagree on the most effective use of current effort and resources. The ability to postpone the satisfaction of current desires in order to protect against global climate disruption and other environmental concerns is nonetheless a crucial element of the process of fostering Green IT.

Broadening Space

In addition to helping us think with a thousand-year time window or longer, Green IT can also help enable scaling of our impact across the spatial dimension. In order to address the world's environmental problems, we will need to work together across the globe to find suitable resolutions. Invariably, different individuals and organizations will bring different experiences, opportunities, and expectations to the discussion. Enabling disparate groups to work together is one of the most important goals that can facilitate sustainability on a large scale.

A global perspective is critical to addressing sustainability effectively. A distinction often made in discussions of international themes involves the separation between the "developed world" and the "developing world." This second term implies a movement toward a higher quality of life and likely greater environmental impact. Perhaps a separation into "more resource-intensive nations" and "less resource-intensive nations," which would create a distinction along somewhat different lines, would be more useful to the current discussion.

It is crucial that people in nations that are currently less resource intensive become engaged with the process of achieving environmental sustainability. For example, over the last twenty-five years, China has accounted for 80 percent of the increase in industrial energy demand (International Energy Agency, 2007); enabling the Chinese and other nations to continue to improve their standard of living without repeating the environmental mistakes of previously industrialized countries is an essential piece of a global sustainability effort. This stance is not intended to align with the hypocritical perspective that suggests that it was acceptable in the past for industrialized nations to pollute on their way to a high quality of life, but that it is not acceptable now for countries aspiring to that same quality of life. Rather, the goal is to help find ways that less resource-intensive nations can achieve a high quality of life for their citizens that is less costly and more sustainable because it takes advantage of recent discoveries and innovations.

One of the advantages that most of the less resource-intensive nations have with regard to sustainability is a lack of entrenched infrastructure that perpetuates environmentally harmful behavior. People in many of these countries are receiving their first introduction to computers through mobile phones rather than desktop computers. If there were to be sustainable systems that could be viable on a large scale, it may be that nations in the process of establishing a modern infrastructure could lead nations with entrenched legacy systems. This phenomenon, known as leapfrogging, can potentially be a catalyst for transformations in the way that the world's civilizations approach environmental concerns.

An ongoing question in the internationalization of the world is to what extent homogenization is a good thing. Certainly, a great deal of standardization has occurred as the world's civilizations have come into closer contact with each other. From passport systems to TCP/IP networking protocols, nations have come to agree on certain ways of facilitating communication and commerce. Digital tools have played a key role in the development of global communication protocols. One might postulate that the homogenization of the world's civilizations might simplify some of the processes of environmental sustainability.

As will be discussed in chapter 2, however, monoculture leaves a farm open to catastrophic failure, while diversity breeds robustness. Although it may be marginally more efficient to have all civilizations work the same way, embracing the differences among cultures will help to develop a global system that is sustainable and robust, and perhaps better able to survive changing circumstances.

The global human rights movement provides a good case study of an effort to enable global cooperation without trying to force unnecessary standardization. The United Nations has long upheld the Universal Declaration of Human Rights (UN General Assembly, 1948), listing rights to freedom, education, and other topics seen as universally due to all humans. This declaration provides a framework for understanding how governments ought to act toward their citizens.

Environmental issues and human rights are closely connected. It is hard to convince people to care about a rain forest fifty years in the future when their children are hungry right now. There is also an inverse correlation between the education level of women and their fertility rate (Akmam, 2002); in this light, the work being done to increase levels of education, in particular among women, may be among the efforts with the longest reach in terms of both environmental and humanitarian effects by slowing the growth of the world's human population. These efforts could work in concert with projects to reduce our per capita footprint to reduce humanity's overall environmental footprint.

In addition to the conceptual relationship between human rights and environmental issues, there is also a similarity in terms of the practical action that is taken, and the way IT factors into that process. For example, in the wake of the 2004 Asian tsunami, over US$7 billion in aid came in from around the world (bbc.co.uk, 2005), with the World Food Programme providing food to 1.2 million people (wfp.org, 2008). Coordinating a response on this scale required the use of many different forms of IT working in concert. Furthermore, since that time, significant effort has gone into the development of early warning systems and other technologies that might mitigate humanitarian crises and coordinate responses to them. Drawing on this and other examples of the rapid mobilization of global resources in a period of just a few months provides an example of how humanity might someday respond to a worsening of the world's environmental issues.

Several major international documents explicitly connect environmental issues to human rights, offering a human-centered view on the topic of environmental sustainability. For instance, the UN Millennium Declaration includes the following section, under the heading "Protecting Our Common Environment":

21. We must spare no effort to free all of humanity, and above all our children and grandchildren, from the threat of living on a planet irredeemably spoilt by human activities, and whose resources would no longer be sufficient for their needs.

22. We reaffirm our support for the principles of sustainable development, including those set out in Agenda 21, agreed upon at the United Nations Conference on Environment and Development.

23. We resolve therefore to adopt in all our environmental actions a new ethic of conservation and stewardship an s first steps, we resolve:

• To make every effort to ensure the entry into force of the Kyoto Protocol, preferably by the tenth anniversary of the United Nations Conference on Environment and Development in 2002, and to embark on the required reduction in emissions of greenhouse gases.

• To intensify our collective efforts for the management, conservation and sustainable development of all types of forests.

• To press for the full implementation of the Convention on Biological Diversity and the Convention to Combat Desertification in those Countries Experiencing Serious Drought and/or Desertification, particularly in Africa.

• To stop the unsustainable exploitation of water resources by developing water management strategies at the regional, national and local levels, which promote both equitable access and adequate supplies.

• To intensify cooperation to reduce the number and effects of natural and manmade disasters.

• To ensure free access to information on the human genome sequence. (UN General Assembly, 2000)

The Millennium Ecosystem Assessment conducted by the UN Environment Programme (2005) sought to "assess the consequences of ecosystem change for human well-being and the scientific basis for action needed to enhance the conservation and sustainable use of those systems and their contribution to human well-being." This assessment involved a number of reports aimed at documenting the current state and trends in the world's ecosystems, then analyzing these data, and offering recommendations about ways for the United Nations and other organizations to take action. One of the "Key Messages" in the summary report stated that "even today's technology and knowledge can reduce considerably the human impact on ecosystems. They are unlikely to be deployed fully, however, until ecosystem services cease to be perceived as free and limitless, and their full value is taken into account" (Millennium Ecosystem Assessment Board, 2005).

Achieving global sustainability will require the collective expertise and participation of people around the world. With a set of common principles underlying the global sustainability effort and IT-enabled coordination across cultures, it may be possible to maintain the cultural diversity that abounds throughout the world while still collaborating on common environmental goals. The ideas presented here may be helpful in establishing the role of IT in facilitating cooperation among different individuals, and by extension among different cultures, around environmental themes.

While environmental issues are affecting all civilizations, many of the examples used in this book relate to people and institutions in the industrialized nations, including those in North America, Western Europe, Japan, and Australia. There are a number of reasons for this focus. First, residents of industrialized nations use a great deal more natural resources per capita and produce more waste per capita than residents of nonindustrialized nations (Diamond, 2008). The industrialized nations, and in particular the United States, have the worst environmental track records of any countries. Second, since many people in the nonindustrialized world are striving to achieve a standard of living like that found in the industrialized one, this focus on current industrialized nations may also provide a test bed for sustainability work that may later be applied more effectively and with greater participation in nonindustrialized nations. By exploring sustainability via the industrialized nations first, this book not only pursues the most striking cases of resource use but may also offer models for sustainable lifestyles that can be deployed elsewhere around the world.

Broadening Complexity

The third main area in which Green IT can broaden horizons is with regard to complexity. There are many different forms of complexity, and Green IT can help with many of them. The two that are most related to the coordinated response of humans to environmental concerns are social and disciplinary complexity.

Social Complexity

In a number of different ways, Green IT can help us work at greater degrees of social complexity. Scaling up social complexity can enable people to take action not just as individuals but also via larger groups and social structures. This scaling can cover anything from small groups of friends to local communities to complex international social networks. One of the great values of computational systems is their ability to scale; online social networks provide for a much greater degree of fluidity and speed in forming like-minded groups.

Among the challenges in managing the social complexity of environmental concerns is the need to pass down innovations from generation to generation. Herein lies the value of attitudes toward sustainability and environmental education; while infrastructural advances can help us live lives with a lower impact, it is critical that future generations embrace sustainability as a lifestyle in order for humanity to survive in the long term. Sustainability is more process than product; it is not simply a matter of fixing the problem and then returning to business as usual. We need a dramatic shift in our approach to living in the world. This shift will produce a great deal of complex changes in human social and cultural systems. It may occur through attitude change, but some aspects are deeply entrenched in human cultures, so adoption may only fully occur in younger generations.

Social complexity also involves the organizations that people have created to facilitate living and working together. Corporations and governments act as proxies for people; just as individuals need to lengthen our time horizons, our conglomerates need to do so as well.

It is interesting to note that participatory culture enabled by the Internet is beginning to blur the boundary between individuals and institutions. It may be that a strong presentation of this dichotomy is becoming less necessary as traditional barriers to group action are lowered. IT support for collective action can be a powerful force for helping people enact change by decreasing the complexity of working together at a global scale.

All cultures contend with the need to consider longer-term issues. The Haudenosaunee (Iroquois) people's seventh-generation philosophy states that decisions should be made with an awareness of their impact on people seven generations from now (SixNations.org, 2007). This philosophy was adopted by the leading brand of green cleaning products, Seventh Generation, Inc. The ability to plan ahead, conceiving of likely futures as well as potential alternates, is perhaps the key attribute that differentiates humans from other animals. Embracing this ability, and allowing people and our technologies to help plan for long-term sustainability, is the core focus of Green IT.

Disciplinary Complexity

In addition to cooperation around the globe, Green IT involves collaboration across numerous intellectual disciplines. Green IT is a relatively young field that draws on many existing ideas and theoretical frameworks. It harvests technological innovations from computer science, information science, human-computer interaction, interactive media, and electrical engineering. It incorporates broad motivations and goals from ecology, economics, and environmental studies. Various Green IT projects have also utilized ideas from anthropology, psychology, sociology, education, art, music, theater, science and technology studies, and numerous other fields.

It is important to distinguish the topic of Green IT from one of its nearest conceptual neighbors: environmental technology. Environmental technology explores ways in which the environmental sciences can be applied to conserve natural resources and mitigate human impacts on the world. Green IT focuses on how IT can serve these and related goals. Of particular interest in the field of Green IT are the personal and social elements—the ways in which people can be enabled to enact lifestyle changes to live in more environmentally sound ways. In addition, as will be discussed later, certain aspects of the theoretical underpinnings of IT lend themselves to environmentally preferable lifestyles and thus make it an ideal platform for sustainable change.

Value of This Approach

The EHCC approach described earlier—focused on broadening people's horizons in time, space, and complexity—has several key values. These benefits primarily relate to the development of technologies and the sociotechnical systems in which they are embedded, but ultimately must lead to an improvement in the world's environmental condition in order to have true value.

First, EHCC can provide a structure for analyzing potential responses to particular environmental problems. Designers can engage in thought experiments regarding the ways people have contributed to certain issues, and how we might be enabled through technology to mitigate or reverse our effects. Potential questions include: What is the time/space/complexity scale of the environmental problem? What is the scale of usual human responses to this problem? How can IT narrow the gap between these two scales?

Second, with particular existing technological innovations, the EHCC approach can provide guidance in evaluating and refining the system to allow it to have greater Green impact. By analyzing the ways in which systems alter people's horizons, this approach may help discover direct or indirect environmental impacts of IT systems that were not originally designed to be Green. With systems that are intentionally designed to be Green, an analysis of how they affect people's scales of time, space, and complexity may help to assess each system's success and overall impact.

Finally, EHCC can be used to compare different technologies and help provide an organizing structure. By arranging systems in a taxonomy based on whether they extend people's horizons of time, space, and/or complexity, and the degree to which they do so, it may be possible to find gaps in the taxonomy that point to fertile space for Green innovation.

Structure

Through both theoretical discussions and an examination of specific cases, this book seeks to provide a way of thinking about Green IT. Chapter 2 offers an overview of the broad environmental concerns currently facing the planet. This chapter presents in detail humanity's resource use, waste production, and other effects on the world, and frames the opportunities for IT to make a contribution. Chapter 3 discusses the characteristics of the human audiences for Green IT, looking in particular at how humans and our societies frequently think and act in narrow ways. Chapter 4 focuses on technology in general, IT specifically, and the sociotechnical systems that arise from them, and considers how they can broaden our horizons of time, space, and complexity. Chapter 5 provides a survey of a broad range of Green IT innovations to offer a sense of the diverse ways IT is impacting environmental issues. Chapters 6–8 discusses three specific areas of Green IT in more detail—environmental education, tools for personal change, and systems that support community action—each of which supplies a unique perspective from which to understand Green IT as a field. In each of these chapters, the analyses are framed around the details of specific projects undertaken by my research group. Taken together, considerations of these three projects help to fill in the contours of the EHCC approach to Green IT. To conclude, chapter 9 explains the way in which Green IT can help us transition from the unsustainable present to a more sustainable future. All of these chapters present an array of existing examples of Green IT projects, in order to provide a sense of the breadth of topics involved in this field.

Methodology

Chapters 6–8 of the book include in-depth descriptions of three social Green IT projects—titled EcoRaft, Trackulous, and GreenScanner—that were created by my research group. EcoRaft is an interactive museum exhibit designed to help children learn principles of restoration ecology. Trackulous is a suite of Web-based tools that assist people in tracking, analyzing, and sharing information about their own environmental impact and other personal data. GreenScanner is a system designed for desktops and mobile devices to help people engage in environmentally preferable purchasing by accessing community-generated environmental impact reviews about consumer products. These projects were created using an iterative design methodology, discussed below, and contributed to the development of the EHCC perspective described previously.

Many varying definitions of design have been suggested by researchers and practitioners over the last several decades. In this text, I will use Herbert Simon's (1996) definition that design is the process of "devis[ing] courses of action aimed at changing existing situations into preferred ones" (p. 111). The use of a design methodology helps ground the book in the practical challenges and opportunities presented by real-world IT systems. The theoretical concepts presented here are derived from ongoing efforts to create working systems.

Iterative design, in particular, offers a way of understanding the context of the use and potential benefits of a system more fully. It utilizes a cyclic process of prototyping and evaluation, with each new prototype being informed by the evaluation of the previous version. The process of coming to understand the possibilities and constraints of the system helps flesh out the set of issues that are most important. In many areas of design, including the design of IT systems, iteration is necessary to produce a high-quality end result. As Web usability consultant Jakob Nielsen (1993) suggests, "Because even the best usability experts cannot design perfect user interfaces in a single attempt, interface designers should build a usability-engineering life cycle around the concept of iteration" (p.32).

In each of these three projects, an interdisciplinary team of designers, engineers, artists, and content domain experts worked together to develop each successive prototype. By bringing together experts from a range of backgrounds, the developers sought to avoid the various problems often encountered by homogeneous teams. For example, in the EcoRaft project (chapter 6), having several ecologists on the team helped ensure that virtual creatures inhabiting the interactive simulation of a Costa Rican rain forest exhibited plausible behavioral patterns and interactions. Had this team been composed only of engineers and artists, the virtual ecosystems would have lacked the veracity that the ecologists could help provide.

For the evaluation of each successive prototype, the research teams used mixed methods to assess the viability and impact of the system on its intended audience. The teams employed both qualitative and quantitative methods at each stage, tailoring the evaluation methodology to the particular needs of each project. With the education-focused EcoRaft project, qualitative methods including semistructured interviews and passive observation were used to assess the user experience and learning objectives. With the Trackulous project, designed to enable people to track their personal habits and behavior, quantitative methods such as an analysis of usage patterns from server logs provided the primary source of information, augmented by a design critique by a skilled user-experience expert. With GreenScanner, the community-based environmental impact review project, server logs as well as online surveys were used to assess the usage and impact of the system.

The audiences that interacted with the various prototypes of each project were selected to represent the target population for the final system. With EcoRaft, the creators exhibited the installation at a regional science museum, and brought in groups of children from the local area to test it at several points. Trackulous was distributed via the Facebook social network and other online mechanisms. With GreenScanner, a press release was sent to a variety of environmental and nonenvironmental news sources as a way of gathering a broad user group through which to evaluate the project.

The iterative design methodology described above facilitated the development of each of these projects, helping the group's research address a range of sustainability issues. In later chapters, these projects are described in greater detail and are each connected to the EHCC perspective. Each project is explicitly analyzed to examine how it contributes to the lengthening of people's time horizon for thinking about environmental issues and the ways in which it situates people at the center of this suite of concerns. This methodology provides an adaptable lens through which to examine the breadth of topics that collectively make up the field of Green IT.

Scope of This Work

The challenge of moving humanity toward sustainability is vastly complex; no one text will be able to cover all aspects of this effort. Therefore, this book seeks to address one small part of the broad set of topics that must be integrated to enable sustainable civilizations. IT is a powerful piece of the sustainability puzzle because it offers a multiplier effect on many sociotechnical systems operating around the world. There are many other factors, however, that must contribute to this goal as well.

There are some deep questions related to cause and effect that underlie the issue of global sustainability. For example, Gus Speth (2008), dean of the School of Forestry & Environmental Studies at Yale University and a global leader in preventing environmental degradation, argues that capitalism, with its emphasis on growth and short-term perspectives, lies at the heart of the world's unsustainability. Rob Watson (2009), who founded the LEED green building rating system, echoes this premise: "The timescales on which markets and nature work are several orders of magnitude off, and most markets and economists continue to labor under the delusion that capital is infinitely substitutable for everything else, including 'ecosystem services.'" Other scholars—for example, Russell Hopfenberg (2003), have maintained that the agricultural revolution may underlie the problem. Researchers from vastly different disciplines, from biochemistry (Elliott & O'Connor, 2007) to public policy (Mazmanian, 2009), all have critical contributions to make in the effort to help humanity live more sustainably.

It is important to explore all of these topics in order to know how best to proceed toward sustainability on a global level; nevertheless, in-depth examinations are beyond the scope of this document. Regardless of the outcome of these and other key debates, it is likely that IT will play a significant role in the solution. IT is not tied to specific food-gathering processes or sociopolitical structures; the concepts presented here can make a contribution in any of the global scenarios that unfold.

Central Argument

Green IT is a growing field that seeks to understand the ways in which IT and environmental issues interact. It takes an interdisciplinary approach, drawing on disciplines from computer science to ecology to economics. There are many projects already in existence in this area, and many more are likely to be developed in the coming years. This book seeks to provide a structure for understanding this field, focusing on the interactions that Green IT systems encourage between humans and the world around us.

The core argument is this:

• Humans are facing significant environmental issues that occur over long time scales, large distances, and great degrees of complexity.

• Unassisted, humans are not well equipped to deal with problems on these scales of time, space, and complexity.

• Throughout history, technological innovations have enabled human cultures to deal with broader suites of problems than we would otherwise be able to address.

• IT specifically involves tools and techniques for dealing with vast bodies of information across wide ranges of time, space, and complexity, and is thus well suited for addressing environmental concerns.

By enabling innovations in infrastructure, education, personal change, community involvement, and many other domains, Green IT can help people live more sustainable lives. This shift need not be instantaneous; rather, we can seek to transition smoothly from current ways of living to more sustainable ones. Green IT systems can assist with all stages of this transition. Once humanity is living sustainably, perhaps we can begin to repair some of the global problems we have caused, restoring endangered ecosystems and threatened species.

The goal of Green IT is to help people make a difference, and change our world to be more in accord with the way we would like it to be. The naked cyclists in Seattle have one unique way of making a difference. In addition to high-profile demonstrations, though, there are thousands of ways, large and small, that people and institutions can participate in a movement toward environmental sustainability. By helping people and organizations discover approaches to changing our world and working together to accomplish sustainable goals, Green IT is enabling a certain part of human consciousness that looks out for others, looks out for animals and plants, and in the long run looks out for our children and ourselves. This text seeks to help us understand the ways in which Green IT has had an impact already, and the ways in which it could have an even greater impact in the future. By contributing to an understanding of the social and technological mechanisms by which Green IT contributes to sustainable living, *Greening through IT* may itself make a greater contribution than the environmental cost of the paper on which it is printed. I hope it will.

2 Environmental Horizons

Unmitigated climate change would, in the long term, be likely to exceed the capacity of natural, managed and human systems to adapt.
—IPCC, 2007a, p. 73

Over the past several years, the popular press has focused a great deal of attention on the problem of global climatic disruption. Climate change has become emblematic of the worldwide environmental movement. Numerous other issues are also significant on a global scale, such as the greatly increased rate of species extinction and the pollution of the oceans. These problems share two important characteristics. First, they have been caused predominantly by humans. Second, each of them may profoundly transform the planet, possibly causing a great deal of death and destruction in the process.

This book deals with the overarching theme that humans are transforming Earth, thereby potentially rendering it unfit for the continued existence of humans and many other species. This transformation is not especially noticeable on the time scale of a human day; however, it is strikingly obvious when viewed from a time scale of decades, centuries, or longer. These broad-scale phenomena point clearly to a set of looming issues that are likely to alter life on Earth profoundly.

To understand how Green IT impacts the world's environmental issues, we must have a sense of where things stand, environmentally. Many other sources of environmental information provide a more comprehensive treatment of this information, especially IPCC (2007a). This chapter briefly covers a variety of environmental concerns that are relevant in the remainder of the book, including population growth, resource consumption, waste production, species extinction, and global climatic disruption.

Population Growth

Perhaps the most critical concern currently facing the effort to limit humanity's impact on the nonhuman organisms and abiotic aspects of Earth is the rapid rate of human

population growth. Interestingly, population size is not always seen as an environmental issue, perhaps because many ideas for reducing human population run afoul of
various ideologies more readily than other environmental topics. Nevertheless, population size has a profound effect on most or all anthropogenic environmental challenges.

As of 19:54 GMT on January 20, 2009, the world population was estimated to be
6,755,118,647 by the U.S. Census (2009b). This number is about 100 million greater
than it was one year previously. In fact, with the exception of a few plagues and other
catastrophes, human populations have been growing continuously for several thousand years, and growing especially rapidly in the past several hundred (UN Secretariat,
1999) (see figure 2.1).

Nevertheless, the rate of growth of the world population has been decreasing over
the last several decades, meaning that we may be reaching a population plateau. A UN
report (2004) estimated that the world population will reach 8.9 billion by 2050 (see
figure 2.2) and peak at 9.22 billion around 2075 (United Nations, 2004).

Looked at close-up over the few decades shown in figure 2.2, it appears that the
plateau might be sustainable, coming gently to rest at a stable population. There is another, less positive possibility, however. To discuss this possibility, it is helpful to go to
an introductory ecology lesson that I was taught as an undergraduate.

Imagine a colony of rabbits that has a relatively stable relationship with the local
population of coyotes. Each coyote eats a certain number of rabbits each year, and
the numbers of coyotes and rabbits settle into equilibrium levels. Now imagine that

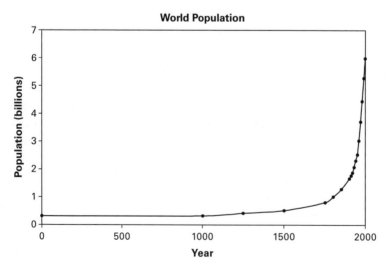

Figure 2.1
The world population over the past two thousand years. From data in UN Secretariat, 1999.

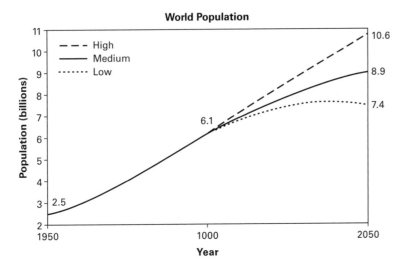

Figure 2.2
Forecasts of human population. From *World Population to 2300* © United Nations, 2004. Reproduced with permission.

the coyotes are removed. Suddenly the rabbit population starts to grow unchecked. Because of their rapid reproduction, before long the rabbits have a significantly larger population than can be sustained by their food supply. They overeat the grass, chewing it down to the roots, and it cannot regrow. Then the large population of rabbits starves to death, except for the few who migrate away or find the remaining bits of grass.

A real-world example of this phenomenon occurred among reindeer on an island off the coast of Alaska in the middle of the twentieth century (see figure 2.3). In 1944, the U.S. Coast Guard released twenty-nine reindeer on the island. By 1963, the twenty-nine had reproduced vigorously; six thousand reindeer now lived on the island. The reindeer ate all the available food; when there was a cold winter in 1963–64, nearly the entire herd died. In 1966, a count showed only forty-two reindeer there (Klein, 1968).

A similar phenomenon could also be happening to people. We have developed techniques for utilizing the world's raw materials and producing food that enable us to have a large number of people on Earth. At the same time, it is not clear that these techniques represent a lifestyle that is sustainable in the long term. It may be that people run out of raw materials, and the system can no longer support our population, leading to a significant reduction in the human population. Understanding and avoiding this potential outcome is an important objective of this book.

While many human-caused environmental issues scale according to the number of people involved, they may become problematic only when the number of people

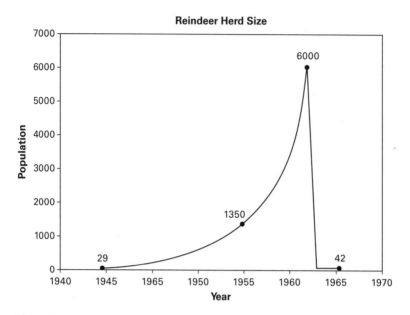

Figure 2.3
Assumed population of the St. Matthew Island reindeer herd. Actual counts are indicated on the population curve. Reproduced with permission from (Klein, 1968).

involved goes above some critical level. For example, if all the plants on Earth can convert a certain amount of carbon dioxide to oxygen, then humans creating carbon dioxide only becomes profoundly problematic when we produce more of it than the plants can recycle, or when we kill off the plants until the equation no longer balances in our favor. Therefore, human population itself is only part of the problem; the way each of us lives provides a multiplier on the population.

Given the materials needed to sustain human life, a finite number of humans can live on this planet. The Earth's precise carrying capacity has been a topic of some debate, with estimates ranging widely based on many different factors (Cohen, 1995). Regardless of the exact number, the more people on Earth, the greater our impact.

Resource Consumption

In order to live, humans consume certain resources and produce various wastes. Some of the resources most relevant to Green IT include space, energy, and raw materials. Each of these resources is discussed below.

Space

Each person needs physical space in which to live. Currently, as noted above, the planet has a population a bit above 6.7 billion people. Earth's area is approximately 200 million square miles (including both land and sea). These figures reveal that there are 33.5 people per square mile, or 832,000 square feet per person. People may exist in a small geographic footprint; in 1990, New York City had 23,700 people per square mile (U.S. Census, 1998), or less than 1,200 square feet per person. If the world were covered with people at the density of New York City, the human population would be 4.6 trillion. These numbers would make it seem that physical space is not a significant limiting factor in terms of achieving sustainability. These statistics, though, do not convey the support structures that are required to maintain a population at such a density. Without farmland, oceans, and other geographic features, there would not be sufficient food to keep the city dwellers alive. In addition, not all land is equally productive when used to produce human food; a square mile of the Sahara desert would make quite a bit less corn than a square mile of Iowan plain.

The distribution of people makes a profound difference to the ecologies of other organisms on the planet. Certain other organisms need much larger habitats to continue to exist. For example, wolves and deer lack the social infrastructure to transport and store food in the way that humans or ants do. For these organisms to maintain breeding populations, large continuous expanses of habitat are necessary. Human roads and other transformations of the landscape have effectively partitioned much of the wild habitat of many species, endangering the stability of those species. As of 1991, in the forty-eight states of the continental United States, there was no location more than twenty-one miles from a road (Foreman, 1991).

Recently, more effort has been exerted to ensure the existence of continuous wild spaces. Wildlife corridors, strips of land linking wild spaces, are used to provide connectivity between the populations in each space. The Yellowstone-to-Yukon Conservation Initiative is seeking to create a continuous wild space from Yellowstone National Park to the Peel River in the northern Yukon, a distance of nearly two thousand miles. The Great Park currently being developed in Orange County, California, includes a wildlife corridor to connect adjacent wild areas. Wildlife corridors improve the survival of species by enabling gene flow between different populations.

More efficient use of space can be made possible by computational modeling of the various human and nonhuman factors of land use (see, for example, [Berry, Hazen, MacIntyre, & Flamm, 1996]). By taking these factors into account, land-use planning can be done more effectively. Determining how heavily to weight the needs of different species will be a source of some contention. How heavily should human needs be weighted in comparison to those of other organisms? How will competing human needs be reconciled? This planning is not being done in a void; rather, it is necessary to take into account the current uses of the land as well.

Energy

A second major resource needed by humans is energy. Currently a great deal of the energy used in industrialized nations comes from fossil fuels such as oil, coal, and natural gas, although nonindustrialized nations still use a great deal of biomass fuel. According to a report by the UK Parliamentary Office of Science and Technology (2002), "Around half of all people in developing countries are dependent for fuel on wood, dung and crop residue, collectively known as 'traditional biomass'" (p. 1). Saul Griffith notes that humanity currently consumes approximately eighteen terawatts of energy from all sources (treehuggertv, 2009).

A large amount of the energy we harvest from the world around us is lost in the process of energy generation (Kaiper, 2004) (see figure 2.4). The figure below shows a graph of the flows of energy used by the United States. More than half is lost in various ways, rather than being converted into useful energy. As the UK Parliamentary Office of Science and Technology (2002) reports, traditional biomass fuel sources have efficiencies of 10 percent, significantly lower than the levels for fossil fuels.

Around the world, oil accounts for 36 percent of the energy supply (U.S. Department of Energy, 2006). In the industrialized world, oil accounts for an even higher percentage of energy consumption. Oil is produced when ancient organic matter is compressed and heated under Earth's surface, and is thus nonrenewable (at least on human time scales).

In a poignant book title, the world's remaining oil reserves have been called "The Last Hours of Ancient Sunlight" (Hartmann, 2004), based on oil's previous life as the sun's energy stored by plant photosynthesis. Given limitations on oil supplies, it is likely that we will soon reach a point where petroleum extraction peaks, and the rate at which we can extract it will begin to decline. The point at the top of the curve, called peak oil (Foucher, 2009) (see figure 2.5), as mentioned earlier, is seen as an important moment in understanding the history of human power consumption; the point at which the global supply of one of humanity's main power sources begins to decline could trigger a major shift in the dynamics of the world's economies. Optimists could see the imminence of peak oil as a good time to begin shifting to other power sources. Pessimists have proposed that peak oil could lead to global economic collapse and significant human population reduction, or other undesirable outcomes.

U.S. Energy Flow Trends – 2002
Net Primary Resource Consumption ~97 Quads

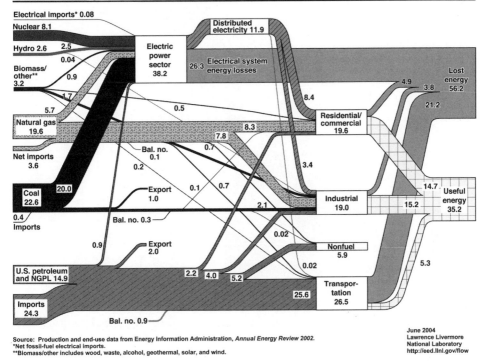

Source: Production and end-use data from Energy Information Administration, *Annual Energy Review 2002*.
*Net fossil-fuel electrical imports.
**Biomass/other includes wood, waste, alcohol, geothermal, solar, and wind.

June 2004
Lawrence Livermore
National Laboratory
http://eed.llnl.gov/flow

Figure 2.4
The flows of energy in the United States. Across all sources, more than half of all the energy in the United States is lost at some point in the process (Kaiper, 2004). Reproduced with permission. © Lawrence Livermore National Laboratory.

Nuclear energy has great potential as an energy source, but has a rocky relationship with environmental concerns. France derives 77 percent of its electricity from nuclear sources (Erlanger, 2008); nuclear energy provides a nontrivial portion of many other countries' electricity needs as well. Nevertheless, while it does not produce the same abundance of pollution as fossil fuels do, nuclear energy leads to nontrivial problems through the production of radioactive waste and the release of waste heat, and profound environmental transformations resulting from accidents such as the 1986 Chernobyl incident. The potential for reactors to produce weapons-grade waste presents additional concerns (von Hippel, 2008).

There has been an ongoing effort around the world over the last several years to seek renewable sources of energy, such as solar, hydro, wind, geothermal, and tidal power.

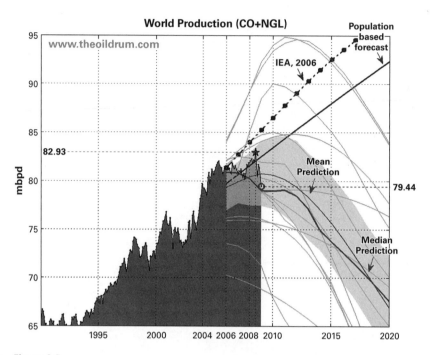

Figure 2.5
Various forecasts of world oil production, most predicting peak oil between 2004 and 2014 (Foucher, 2009). (cc) Sam Foucher.

There is great potential in these sources; for example, eighty-five thousand terawatts of solar energy hit Earth, and there are thirty-six hundred terawatts of wind (treehuggertv, 2009). Some countries already draw a large percentage of their power from renewable sources; Iceland, for instance, in 2004 drew 78.4 percent of its electricity from hydropower, and another 13.7 percent from geothermal (Orkustofnun, 2005). Moving to sustainable energy sources may still provoke a variety of environmental concerns. These objections may be practical, such as excess wind power at certain times causing dams to need to release more water into rivers, thereby harming the salmon living there (Tilkin & KATU Web Staff, 2008), or aesthetic, such as opposition to "ugly" fields of windmills (Griscom, 2002). Nevertheless, renewable power is often seen as an important component of the move toward global sustainability.

Raw Materials
Humans consume a great deal of raw materials. Some of these—such as oil, coal, and natural gas—are used as energy sources, as described above. Many other natural re-

sources, though, are used for enabling human sustenance, energy transfer, transportation, communication, health care, knowledge archiving, entertainment, beautification, and many other purposes.

For example, since humans and many other animals need oxygen to breathe, clean air is another critical resource in this analysis. Air is currently not as limited a resource as many of the others on this list, but it is involved with a number of environmental issues, mostly as a result of pollution and human waste production, which will be dealt with in the next section. Various researchers around the world are studying air quality, from the Air Quality Research Center at the University of California, Davis to the Clean Air Initiative: Asia Partnership.

Water is another key resource, used for an assortment of purposes, from drinking to irrigating crops to cooling power plants. Enabling humanity to continue to have a steady supply of clean water is essential to our continued survival. The study of hydrology has been growing as an academic field as this need has become more apparent. The Urban Water Research Center at the University of California, Irvine is one of the leading centers in this field.

Food is also a central issue in many environmental discussions. The United Nations has a Food and Agriculture Organization tasked with "helping to build a food-secure world for present and future generations" (Diouf, 2001). The effort to produce food for 6.7 billion people is challenging, especially given the vagaries of climate; often, civilizations overproduce in order to guarantee that there will be sufficient food for any inhospitable times that might arise.

A challenge of the environmental movement is the possibility that it conflicts directly with one of the foundational premises of the dominant global civilizations. Various researchers and authors (Hopfenberg, 2003; Quinn, 1997) have argued that totalitarian agriculture—that is, the agricultural style practiced by most of the world's communities in which all food belongs to humans, to the exclusion of all other species—lies at the root of the current global environmental crisis. Daniel Quinn proposes that totalitarian agriculture is not inextricably linked to humanity by its very nature. Rather, people lived without it for hundreds of thousands of years. About ten thousand years ago, people invented totalitarian agriculture. Since that time it has spread rapidly through human cultures and enabled cultures that practiced it to grow quickly. As such, it is one of the shared premises among most human civilizations. Quinn contends that it is both possible for humans to live without it, and necessary in order to prevent the impending catastrophe of human overpopulation, the extinction of most other species, and then a vast human population reduction.

Beyond the use of energy sources, air, water, and food, human civilizations consume many other raw materials, from wood and iron for building and construction, to chemicals and ores for manufacturing, to precious metals and gems for adornment. The extraction, transport, and transformation of these materials into usable products lead to a great deal of energy consumption and secondary resource use. In addition, these human processes lead to much waste production, which is the focus of the next section.

Waste Production and Disposal

There are numerous waste materials produced by the diversity of human activities. This section addresses them in three main categories: human waste, material waste, and energetic waste. These various forms of waste all impact human lives; as William McDonough and Michael Braungart (2002) offer, referencing an often-used water metaphor for the process of resource consumption and waste production, "In planetary terms, we're all downstream" (p. 127). Beyond supplying the raw materials of human life, a sustainable society needs to provide for the processing and/or removal of waste in a humane way. Green IT can offer support for addressing many problems associated with humanity's waste.

Human Waste

The biological process of human life produces several main by-products: CO_2, feces, and urine, not to mention dust, moisture, heat, and an assortment of other leftovers.

Human beings produce a great deal of CO_2 as a result of our respiration process. According to data from the Carbon Dioxide Information Analysis Center, each person generates about 1 kilogram of CO_2 per day (cdiac.ornl.gov, 2009). This amount means that the world population generates approximately 2.4 trillion kilograms of CO_2 over the course of a year, or over four times more CO_2 than is emitted by the aircraft industry (Guardian News & Media, 2008). On cursory analysis, one might imagine that humans are adding a huge amount of CO_2 to the atmosphere. However, this CO_2 is part of a closed loop between people and the organisms we eat (cdiac.ornl.gov, 2009) and does not contribute significant GHG to atmospheric concentrations. Nevertheless, beyond the CO_2 produced by people breathing, there are many other ways that we cause CO_2 to be produced; these forms of waste will be dealt with in the section on "Material Waste" that follows.

The average American will produce approximately 550 kilograms of urine and feces over the course of a year (Pimentel et al., 1980). With 6.7 billion people, that level of excretion would amount to 3.6 trillion kilograms of excreta annually. (It is likely that civilizations with lower levels of consumption would produce less bodily waste; as many societies move toward higher-consumption lifestyles, their bodily waste will likely increase as well.) This material contains bacteria that can propagate illnesses and may be a pathway for pharmaceuticals to have unintended negative effects (Donn, Mendoza, & Pritchard, 2008). Conversely, human waste is nutrient rich and can be reused in an assortment of ways. Enabling this waste material to contribute again to society, rather than being a burden, requires some form of treatment or recycling.

Humans also shed skin cells (one of the main components of household dust), emit moisture through perspiration, and give off heat. Moreover, while it may seem a bit macabre, a non-negligible by-product of human civilization is dead bodies. With a population of 6.7 billion, an average life span of 67 years (World Bank, 2001), and an average weight between 60 and 70 kilograms (Taylor et al., 2006), human civilization produces on the order of 6.5 billion kilograms of dead bodies every year.

Material Waste

In addition to the waste that we produce as a result of our basic physiology, we produce a great deal of waste because of our lifestyles as well. A critical environmental issue around the world is the great deal of waste produced from the production and consumption of consumer products. As mentioned earlier, the United States produces 230 billion kilograms of municipal solid waste per year (U.S. Environmental Protection Agency, 2008a). In order to enable the world's civilizations to exist in their current form, corporations and other organizations produce a large amount of industrial waste.

Besides the production of solid waste, air pollution has been a global environmental problem for centuries, from the burning of wood around the world (apparently it soiled the togas of the Romans [Bregman & Lenormand, 1966]) to modern issues relating to chlorofluorocarbons and other gaseous pollutants.

CO_2, which was discussed briefly above, has become one of the central human waste products in many environmental discussions due to its impact on global climatic disruption. Many different processes produce CO_2, in particular the burning of fossil fuels. Since so many human actions require the use of energy in one form or another (electricity, fuel for transportation, etc.), CO_2 is an important pollutant, currently in the foreground of many people's awareness, with 6.5 million hits on the Google search engine for the terms "CO2 'Global Warming.'"

The flows of materials in human societies have profound environmental impacts, but these impacts are often invisible to or incomprehensible by the people taking action in those societies. For instance, many people keep a trash can next to their toilet. When they have used a tissue, they might throw it into the trash or the toilet. This small choice causes the tissue to go on profoundly different paths. The tissue that went into the trash probably gets tied up in a bag, put in a dumpster, loaded onto a truck, and dumped in a landfill with vast quantities of other solid waste. The tissue that went into the toilet got waterlogged and flushed, flowed through the sewers, disintegrated, and ended up in a sewage treatment plant, where it was processed with vast quantities of other liquid waste. Which of these paths is more environmentally sound? How would one determine the difference? Presumably the difference is measurable, but it is certainly more effort than it is worth for most people to measure it. The difference is much greater, however, when multiplied by three hundred million U.S. residents or billions of people around the world. At this scale, it may well be worth the time for someone to investigate. Providing the IT tools to help people study such issues and propagate the results to a broad audience could be transformative for humanity's collective progress toward sustainability.

Energetic Waste (Heat)

A significant by-product of many energy-consuming undertakings such as manufacturing is a great deal of heat. Heat in itself does not seem particularly dire; still, temperature changes can have profound impacts on certain ecosystems. The challenge with heat is moving it from place to place. Water is often used for transporting heat. This heated water may then cause environmental problems. Stream and river habitats downstream from nuclear power plants, for example, are sometimes disrupted by the heated water that is returned to a river. Recapturing the heat that is produced by various human processes is one of the challenges facing a sustainable civilization.

Effects on Other Species

Beyond the direct resource consumption and waste production caused by human civilizations, people also have profound effects on the other species around us. This impact can take a wide variety of forms, from the vast success that cows, dogs, and oil palms (see figure 2.6) have enjoyed as a result of symbioses with humans, to the complete eradication that people are causing, intentionally or unintentionally, for thousands of species every year.

Symbioses and Monocultures

Over the course of our own great success around the globe, humans have enabled a variety of other organisms to thrive as well. We have enabled vast numbers of cows, pigs, and chickens to be born, and wide fields of corn, soybeans, and rice to grow, to

Figure 2.6
An oil palm plantation (Dittmar, 2008). © Percita Dittmar.

provide food supplies. We have millions of dogs, cats, and other animals living as pets. Many human houses and yards feature nonnative philodendrons, palms, boxwoods, and other ornamental plants. We unintentionally bring rats, pigeons, cockroaches, bacteria, and viruses with us in most of our colonization efforts. Through these and other mechanisms, we encourage the growth of certain species well beyond the success that they might have enjoyed had humans not proliferated so abundantly.

A challenge that this vast success for those species creates, though, relates to the practice of monoculture. Monoculture is an agricultural practice in which a single crop is grown over a wide area. Monocultures provide a distinct economic advantage by enabling the standardization of planting, caring for, and harvesting that species. However, monocultures are often problematic as well. For example, they can lead to brittleness, since a single pathogen can sweep rapidly through the entire population due to the homogeneity and close proximity of the organisms. Any environmental change that moves growing conditions beyond the range for a single individual effectively moves them beyond the range for the entire population. Perhaps the most famous instance of the collapse of a monoculture is the Great Irish Famine, mentioned earlier, in which more than a million people died due to a potato blight in the mid-nineteenth century. Some of the principles of monoculture apply to other dense populations too; since species that benefit from human civilizations tend to do well (since humans have done well), the concentration of so many resources in a small number of species can lead to a lack of robustness.

Species Extinction

While some species thrive as a result of human activity, most do not. Ecologists have observed that the current rate of species extinction has earned the modern era a place in geologic history as the sixth great extinction event since the beginning of life on Earth.

In broad terms, extinction is a perfectly viable piece of the evolutionary process. Species have been going extinct since shortly after the beginning of life (Wilson, 1992). The current rate of extinction, thirty to fifty thousand species per year (Leakey & Lewin, 1996; Olson, 2005), though, greatly exceeds the baseline rate over most of the history of life. As noted paleoanthropologist Richard Leakey and anthropologist Roger Lewin (1996) write, regarding the vastly increased rate of extinction currently occurring around the world:

Even if we take a figure in the lower range of estimates, say thirty-thousand species per year, the implication is still startling. David Raup has calculated from the fossil record that during periods of normal, or background, extinction, species loss occurs at an average of one every four years. Extinction at the rate of thirty-thousand a year, therefore, is elevated 120,000 times above background. This is easily comparable with the Big Five biological crises of geological history, except that this one is not being caused by global temperature change, regression of sea level, or asteroid impact. It is being caused by one of Earth's inhabitants. *Homo sapiens* is poised to become the greatest catastrophic agent since a giant asteroid collided with the Earth sixty-five million years ago, wiping out half the world's species in a geological instant. (p. 241)

Aside from the deaths of the organisms themselves, there are many other problems that go along with mass extinction on a global scale. First, when many species go extinct over a relatively short period of time, the ecologies of many other species are transformed profoundly as well. For example, mass extinction due to climate change may then cascade through other species. A species' food source or pollinator may disappear, resulting in population reduction or extinction, or that species' predator may disappear, resulting in brief and catastrophic overpopulation. Rapid change may also enable more sustained success of a given species, as it expands to fill newly available niches. Nevertheless, the transformation by which those niches become available would be a time of great uncertainty for the continued existence of any given species. Niles Eldridge and Steven Jay Gould (1972) termed this model of evolution—long periods of evolutionary stasis, followed by rapid change—"punctuated equilibria." Such a period of instability could be particularly problematic for humans, as there are many species that we rely on for our survival, and the extinction of one of the big ones (for example, corn, wheat, or rice) could compromise the nutritional infrastructure on which human civilizations rest.

Second, a reduction in species diversity causes a lack of robustness with regard to cataclysmic events. In a world with many species, any given global disaster is unlikely to eliminate all species. In a complex ecosystem, some organism is likely to have some adaptation that will enable it to survive. By causing the rapid elimination of so many species, humans may be compromising the world's robustness—that is, the biosphere's ability to withstand stresses such as global climatic disruption, or an incident such as a meteor strike or nuclear war, without profound transformation.

Effectively, we are beginning to create a monoculture of humans. While humans are perhaps more adaptable than wheat or potatoes, we do run the risk of creating a brittle ecosystem by enabling the elimination of species at such a rapid rate.

Global Climatic Disruption

Perhaps the most commonly discussed environmental challenge of the modern era is global climatic disruption. The IPCC (2007b) "Summary for Policymakers" report offers that "warming of the climate system is unequivocal, as is now evident from observations of increases in global average air and ocean temperatures, widespread melting of snow and ice, and rising global average sea level" (p. 5). The report then goes on to state that "most of the observed increase in global average temperatures since the mid-20th century is very likely due to the observed increase in anthropogenic greenhouse gas concentrations" (p. 10).

Climate change is problematic because it may cause the habitats of many living species to change sufficiently that they can no longer survive. In addition, it may lead to large areas of land going under water as sea levels rise. Through both of these effects, global climatic disruption is likely to exacerbate the significant extinction trend described above.

The Unfortunate Possibility of Mass Extinction

Looking forward, a wide array of possibilities remains for how the world will be impacted by environmental change. The goal of this book is to provide a hopeful vision of a sustainable future, well grounded in strong science. However, in order to appreciate the need for changes such as those described here, it is relevant to visit briefly with a devastating yet not implausible scenario for the future of Earth.

Here is how a doomsday scenario for the future might unfold. For the next fifty years, human population continues to grow, and the wild areas of the world are mostly cultivated, causing a 90 percent reduction in the world's biodiversity. The vast majority of the planet's biomass accumulates in a relatively small number of taxonomic groups—humans, corn, wheat, soybeans, cows, chickens, farmed salmon, cockroaches, ants, pigeons, rats, bacteria, and mycorrhizal fungi. On a day much like today, a certain wind-borne parasite undergoes a mutation, making it slightly more efficient in exploiting cornstalks as a host. Within a few months, the global corn market has collapsed, causing famine and war on a worldwide scale. As people in rural areas turn their efforts to personal survival, support services to the cities cease to operate effectively, and most of the people in urban areas die of dehydration, starvation, or disease within several months. Some people on rural farms survive longer, especially those individuals more than one tank of gas from a city (the radius within which the city dwellers were able to pillage before starving). Since most of the world's people live in cities, and nearly all of the rest live within a 400-mile radius of cities, this scenario means that the vast majority of the people in the world would have died.

Instead, a (somewhat simplistic) best-case scenario goes something like this. A new company develops an extensible solar panel and battery system that efficiently converts sunlight to electricity, stores it effectively, and scales well; it captures 90 percent of the global energy market by 2040, making its inventor the richest person in the world. The advent of low-cost, sustainable, pollution-free energy increases the quality of life for most of the world's people. The globe's population levels off at nine billion and then starts to decline slowly. The solar energy also helps to make large-scale recycling feasible in a way that it had not been previously, and makes it financially viable to convert landfills back into usable material. New technologies, sustainably designed and designed to enable sustainability, help people live well with smaller environmental footprints. Since human populations are declining, we no longer need to colonize new areas of land, thereby greatly reducing the rate at which species are being lost. Simultaneously, restoration efforts enable endangered and threatened species to come back from the brink. The abundance of energy allows humans' quality of life to be high and facilitates effective recycling on a large scale.

The reality of the situation probably lies somewhere between the two scenarios described above. In the absence of a "magic" power source that will miraculously solve all the world's issues, the process will be more difficult, requiring effort on many fronts by many people and organizations. Earth is undergoing vast changes as a result of human expansion. Unchecked, these changes threaten to undermine the ecological foundation on which human civilizations are built. If the issues are presented to the people of the world in a compelling way, however, from both grassroots and governmental levels, it is possible that people may work together to come up with a viable future. Because of the time scale of the issues involved, we are actually coming quite late to a broad realization of the problem. Nevertheless, rapid mobilization on a global scale can address the issue before the transformation leads us to catastrophe. There is a growing trend toward environmental technologies—hybrid cars, green industrial design, and many other efforts—that, when taken together, are a great first step in this process. Ultimately this effort is necessary for the survival of our descendants and the continuation of living things in general.

Humanity is facing significant environmental challenges on many fronts. Human populations, and our resource use and waste consumption, are responsible to a large degree for these challenges. We are creating monocultures of some species, eliminating many others, and changing the climate in ways that are likely to have drastic impacts over the coming century and beyond. These impacts are occurring over long periods of time and across the entire world, and involve a great deal of ecological and social complexity. An understanding of these issues is an important part of launching an effective response; this chapter has sought to establish a broad basis for further discussions of the world's environmental issues and how Green IT may be able to help.

3 Human Horizons

The strong scientific consensus on the causes and risks of climate change stands in stark contrast to widespread confusion and complacency among the public.
—John D. Sterman, 2008, p. 532

The previous chapter sought to describe some of the core environmental challenges currently facing humanity. In addition to understanding the environmental issues themselves, it is essential to understand how and why people engage with these environmental concerns. In order to bring about significant change, large numbers of individuals must alter their behavior. We humans change our behavior in a number of ways. We are willing to exert a great deal of effort in support of our beliefs. We are also willing to change if the infrastructures and policies surrounding us change, and make it easy or necessary for us to change as well. This chapter provides an introduction to some of the human factors most relevant to ideas of sustainability.

One of the great challenges facing the sustainability movement is the fact that, even though people sometimes understand the short-term aspects of the phenomena, we have difficulty comprehending the long time scale nature of environmental issues. This disconnection has roots in both evolutionary biology and cultural factors. This chapter explores both of these topics, and considers how people might be able to move forward with environmental ideas and action, given the current realities of human culture and behavior.

Evolutionary Roots

One of the key aspects of evolution-based animal behavior is the drive for an individual animal to survive. Contributing genetic material to the next generation of organisms entails the individual's not dying before passing on that material. Therefore, selfishness lies at the core of what it means to be an animal and hence a human. While there are many other attributes of what it means to be human as well, at a basic level we are self-serving whenever we breathe, eat, or dress warmly when it is cold outside. Being self-serving has a bad reputation in human societies, primarily because it is usually shorthand for "being self-serving to the disadvantage of another human." Protecting and perpetuating oneself, however, is an important part of many socially condoned acts—returning a favor, caring for one's family, or doing good deeds. In fact, the ultimate act of failing to perpetuate oneself—suicide—is considered illegal or immoral by many social structures around the world as well. Staying alive is well grounded biologically and also viewed as a social good.

Humans have a great deal of cognitive and physiological apparatus that supports the effort to survive—avoiding danger, finding food, attracting mates, and raising children. Natural selection favored humans who did well in these areas, leaving us with a wide range of capabilities: senses such as vision, touch, and hearing; memory of events that tasted good or hurt us in the past; sexual desire; and parental love. While these capabilities certainly do not prescribe all human activity, they nevertheless form the underpinnings for much of our behavior.

There are many different evolutionary biological theories that potentially affect what people care about, and thus the way in which people might be motivated to take environmental action. This section discusses a few of these concepts.

Coefficient of Relatedness

A key aspect of evolutionary theory that appears to be fundamental to what people care about is called the *coefficient of relatedness* (Wright, 1922). This number specifies the level of relationship between two individuals, with identical twins having a value of 1, parents and children having a value of 0.5, and less related individuals having even lower values. Drawing on this premise, the concern that a person might have for another organism's life should increase in proportion to the coefficient of relatedness between the two. As genetic similarity increases, so too does concern. As resources become scarce, though, the slope of the curve would become steeper, with individuals caring less and less about organisms dissimilar to themselves. These ideas could be relevant in trying to answer questions such as whether a million dolphins are worth a single human life. Looking at the coefficient of relatedness and the scarcity of resources can help justify the spirit of the interspecies interactions that are likely to occur between humans and that species. Certain interactions could be of sufficiently minimal value that there is effectively a threshold below which one organism is of no biological concern to the other one. There are also psychological and emotional effects, such as the sadness that accompanies the death of a pet, that may contribute to humans protecting certain nonhumans; nevertheless, the coefficient of relatedness may set a baseline for people caring about other organisms.

This theory is relevant to environmental concerns because it suggests that, in times of scarce resources, we humans think more about preserving ourselves and our nearest relatives. Even in times of abundance, there is still a fairly sharp falloff in terms of this curve; we tend to care more for our families than for other people, and more for other people than for other animals. We have relatively narrow horizons of social complexity.

Cooperation

Beyond the purely genetic incentive to support another individual, humans and many other animals have the ability to work together, thereby enhancing individual fitness. Because of this cooperation, there exists an evolutionary pressure to be able to keep track of social partners with whom one might collaborate—that is, other members of one's social group. Robin Dunbar (1993) has offered that there is a correlation in primate species between relative neocortex size and social group size. That is, primates with small neocortices compared to their total brain size tend to live in small groups, and those with large neocortex ratios live in larger groups. From this correlation, he proposes that the typical size of human groups would be approximately 150 individuals (sometimes called "Dunbar's number"), with a relatively large margin of error (a 95 percent confidence interval of 100–231). The benefits of collaboration could provide an incentive for looking out for one's valued collaborators, thereby changing the degree of concern that one has for various individuals beyond the level dictated by the coefficient of relatedness.

Conspicuous Consumption

Biological researchers have suggested that the phenomenon of *conspicuous consumption* (Zahavi & Zahavi, 1999; Roberts, 1998) can be an evolutionarily viable survival technique.[2] Conspicuous consumption enhances an organism's fitness because it demonstrates that the organism has sufficient resources to live, and then some. This abundance of resources suggests to other members of the organism's community that it may be a valuable social or sexual partner, with sufficient resources to squander some on goals beyond mere survival. The term "squander" will be employed to mean the consumption of resources that could be used for survival on goals that are not directly survival related. Squandering may still be a viable indirect mechanism for surviving and propagating one's genes.

Two forms of conspicuous consumption are particularly notable. *Sexual handicapping* (Zahavi & Zahavi, 1999) involves an individual's exhibiting resource-intensive behavior or morphology in order to communicate his or her (but usually his) good genes. For example, the brightly colored tails of many birds are a significant handicap to those organisms; the tails require a lot of energy to produce (since bright pigments are energy intensive) and make it harder for the organisms to survive (since predators can see them). A bright tail conveys unequivocally that the bird sporting it is a "winner," with sufficiently abundant resources to have lived to maturity despite such an impediment, and is therefore a prime mate.

A second form of conspicuous consumption is *competitive altruism* (Roberts, 1998). In this behavioral pattern, organisms behave in pro-social ways, issuing alarm calls or saving the offspring of other members of their community in order to demonstrate their abundance of resources. Similar to sexual handicapping, competitive altruism is a drain on the individual's resources, and marks that individual as a high-quality social or reproductive partner.

There are several characteristics that make an attribute or behavior a good vehicle for exhibiting conspicuous consumption. It must be obvious, so that other members of the target community can recognize it. It must be accurate; community members must be able to use it to evaluate the relative merits of different individuals. Finally, it must be "unfakeable;" that is, it must be easier for the organism to exhibit the attribute or behavior than to exhibit an indistinguishable facsimile (Zahavi & Zahavi, 1999).

Like many other animal species, humans exhibit a tendency for conspicuous consumption (Veblen, 1912). To an evolutionary biologist, a sports car shares a number of characteristics with a peacock's tail. The bird's tail is obvious; so too are the logos and characteristic body shapes of sports cars. The bird's tail requires the expenditure of substantial resources. So too does the sports car. Finally, the bird cannot attach a fake tail to itself; neither is it economically viable to manufacture a fake sports car. Both the tail and the sports car serve as reliable indicators of resource abundance.

Similarly, people engage in competitive altruism in a range of ways. The high-end grocery store Whole Foods has begun selling a grocery bag with a large logo reading "Feed the Children of the World" on it. To own this bag, a shopper must pay US$29.95, of which US$10 will be donated to the World Food Program's Rwanda School Feeding operation. This amount is sufficient to provide one hundred meals to children in Rwanda. This demonstration of resource abundance may not only make the bearer feel good but it also may cause others around them to consider them worthy social or sexual partners. While one might think that charity might not be relevant to sexual ends, a group of psychologists recently found that "although mating motivation did not lead women to conspicuously consume, it did lead women to spend more publicly on helpful causes" (Griskevicius et al., 2007, p. 85).

Whenever resources are squandered, there must be some "sink" into which those excess resources are poured. For example, when a species of birds tends to have long colorful tails, their resources are being poured into the populations of snakes and other predators that are better able to catch and eat the birds because of their highly visible tails. When people drive BMWs and other luxury cars, the resources are being poured into the corporations that produced them. (When people drive luxury SUVs, the resources used to buy the gas needed to move so much metal around ultimately end up poured into the atmosphere as carbon dioxide and other pollutants.) Snakes and corporations serve the same role in this system; they are the keepers of the "unfakeability" needed to make a trait costly.

Corporations, like snakes, are not evil. They are simply tasked with one purpose—maximizing shareholder value. Unless there is a profound shift in the corporate law in the United States and many other capitalist countries, it is unlikely that environmental sustainability will ever be the primary goal of corporations. Sustainability may soon be an indirect goal for many corporations, when consumers have sustainability as a primary goal and the corporations must satisfy the consumers. Nevertheless, the world could perhaps be made more sustainable if some portion of the human urge to squander resources could be piped into efforts with direct social benefits.

Systems that let us engage in conspicuous consumption (as we appear to need to do in order to demonstrate our quality as social or sexual partners), but do so without wasting resources unnecessarily, could be the basis for new social interaction styles. In an article in an in-flight magazine I read while writing this book, the author discusses his two-week experiment with car sharing, explaining, "My nearby 'shared' car is a Toyota Corolla sedan—nowhere near as cool, or nimble as my Jetta. . . . In the new world of reduced carbon footprints, maybe we'll all have to find ego gratification elsewhere" (Lerner, 2009, p. 80). An understanding of the biological and cultural issues underlying various social and identity-based phenomena could help inform the creation of a range of new technological systems that support novel, more sustainable activities and interactions.

On an evolutionary time scale, while humans were able to control factors such as their personal location (e.g., near food or away from danger), the actions they took (e.g., eating, mating, fleeing, or cooperating), and some characteristics of their environment (e.g., what objects are nearby or whether rain falls on one's head), they were not able to affect large-scale processes intentionally. Without being able to have this effect, there was no evolutionary pressure to select for cognitive or physiological apparatus to manipulate this kind of issue, so we did not evolve any. Luckily, as is discussed in the next section, humans have developed culture, which has broadened our ability to think and act collectively.

Cultural Effects

The abilities to remember, communicate, learn from communication, and apply that learning in new situations enable humans to form elaborate behavioral patterns—collectively called culture—that are not always obviously directed toward satisfying biological imperatives. Culture allows individuals to adapt to changing circumstances in the world much more rapidly than genetic evolution. Richard Dawkins (1990) has suggested that cultural evolution resembles biological evolution, via the transmission of cultural units called "memes" that serve a similar role to genes in biological evolution. While until the last few millennia culture did not have ways of intentionally effecting change at vast scales, it nonetheless provides the scaffolding for the large-scale social responses needed to address our current environmental concerns. Humans can transcend some of the limits imposed by biology and evolution by means of cultural innovation.

Culture can enhance our ability to respond to environmental issues in a number of ways. It can enable collaboration on previously impossible scales (e.g., the United Nations, Wikipedia, etc.). It can support scientific research so we can understand how the future is likely to unfold. Perhaps most important, it can give people the freedom to take action based on our best understanding of how to protect the future of our species and all living things.

One of the challenges of encouraging people to act in environmentally sound ways is helping us see that we are part of nature, rather than separate from it. People often use the term "nature" to refer to organisms that are not part of human civilizations—that is, organisms that are not domesticated animals and plants, nor people. It is potentially problematic, however, to see humans as separate from nature. There are deep aspects of the Western worldview that contribute to a separation between humans and the ecosystems in which we live. In his article "How Computers Contribute to the Ecological Crisis," C. A. Bowers (1990) offers that the Cartesian mind-set, which places humans squarely at the center of the universe, causes people to see the world's environmental problems as being separate from humanity, rather than human cultures as being integral to the problems. This separation leads to a sense that *we* (humans) need to do something about *it* (nature), rather than us being *part of* it. This perspective contributes to the challenge of helping humans live sustainably, by creating a disconnection between humans and nonhuman species. Bowers suggests renaming the "ecological crisis" as the "ecological/cultural crisis," to draw attention to the central role of human cultures in the issues at hand.

A survey of newspaper headlines demonstrates that humans have the capacity to exhibit concern about environmental issues (Cappiello, 2008; Rosenthal, 2008; Sun, 2008). Many individuals and groups have clear commitments to environmental causes, from politicians to artists to farmers. To provide just one example, the digital artist Vibeke Sorensen (2002) described her work in the following way:

I want to use technology to help us evolve better, that is more ethical, more humanistic, more ecological, more responsible,—more esthetic structures, patterns, and approaches to interacting with the world,—nature as well as other people—that help put us into harmony with them, so that we will be better able to live in and with the world, and help it and us to survive. I consider it urgent, given the rate of destruction of our ecosystems and the move towards war and human suffering. From my point of view, technology is only truly useful if it can do that.

Humans do not always prioritize environmental issues in our decision making, though; often we make decisions based on economics, security, entertainment, religion, and many other factors, even if these issues conflict with environmental preferences. There are many differences among cultures that affect people's ability and interest in responding to environmental change, and provide challenges for efforts to mount large-scale responses. According to a UN Educational, Scientific, and Cultural Organization (UNESCO 2002) educational program, "The dynamic balance between cultural differences and [the] emerging global ethic is a key concept in educating for a sustainable future." This section looks at some of the factors that contribute to and detract from a focus on environmental topics.

Corporations

It could be proposed that capitalism, by its very nature, is problematic from an environmental perspective. Capitalism is based largely on competition, with people working to accumulate wealth in as great abundance as possible. The current model of capitalism entails producing goods as inexpensively as possible (even luxury goods are manufactured as economically as possible while still maintaining high quality), largely without regard to environmental concerns. There are certain governmental restrictions on manufacturing, and certain markets demonstrate a preference for goods produced in an environmentally sound way, but beyond responding to those factors, the central goal is producing goods cheaply and selling them at a premium. In fact, the executives of a publicly traded corporation have a legal responsibility to enhance shareholder value. If they behave in a manner that does not maximize profits (such as choosing a more expensive manufacturing process that has a lower environmental impact), they could be liable for prosecution. Since in many cases more environmentally sound processes are more expensive, this would appear to point to a direct conflict between capitalism and environmentalism. In addition, environmental costs tend to occur further in the future than other costs (Regnier & Tovey, 2007), reinforcing a disconnection between short-term financial gain and longer-term environmental concerns.

Many corporations actively reinforce the consumption-based economies in place around the world. For example, Discover Card runs an advertisement that begins, "We're a nation of consumers . . . and there's nothing wrong with that" (cbs.com, 2008). The business models of these companies are predicated on people consuming more every year.

Nevertheless, the accumulation of wealth is not inherently opposed to the goal of environmental sustainability. Rather, it could be directly in line with it. Corporations such as Ben & Jerry's have made environmental concerns a cornerstone of their corporate image, and have thrived as a result. With an environmentally motivated consumer population, market forces can drive the production of goods toward environmental sustainability. When dealing with a consumer base for which environmental issues are of primary concern, executives of companies could be violating their fiduciary duty by *not* causing their company to act sustainably. The challenge thus becomes the education of the consumer base.

The rise of an empowered community of individuals can force change at a corporate level. For example, the mega-retailer Wal-Mart had been accused of shortcomings in labor relations and environmental responsibility (Kabel, 2006). With the advent of organizations such as WalMartWatch.com and others, though, a new level of oversight became possible. "There was a time where people in business believed all they had to do was run their business," said David D. Glass, ex-CEO of Wal-Mart. "But it doesn't work that way anymore. There is an accountability that goes way beyond that" (Rosenbloom & Barbaro, 2009, p. 2).

Ultimately, sustainability will become more and more connected to economic viability. According to the *Smart 2020* report, former UK government and World Bank chief economist Lord Stern proposed that "to ignore rising carbon emissions that will result in dangerous climate change now will damage economic growth in the future" (Climate Group, 2008, p. 12).

Governments
Governmental regulations and laws provide a mechanism through which groups of individuals have encouraged their members to do things they might not want to do

without external pressure. These organizations seek to change the motivational land-scape on which people act in order to bias our behavior toward certain actions and away from others.

One key challenge that governments often help to address is the "Tragedy of the Commons" (Hardin, 1968), in which a collection of individuals, such as cowherders, acting out of their own self-interest eventually destroy a common resource, such as a piece of land. By regulating the degree and manner of use of shared resources, govern-ments assist in curbing overexploitation. In this way, governments take a broader view than any one individual can, thereby enabling all individuals to benefit in the long run.

Governments tend to deal with scales of time, space, and complexity beyond those with which most individuals concern themselves. They plan for longer periods of time such as decades or centuries. Correspondingly, they often act more slowly than in-dividuals, taking years to change policies. Governments control large areas of land, sometimes covering millions of square miles. They also seek to support the apparatus for meaningful decision making on large scales of complexity, frequently undertak-ing the management of economies, utilities, warfare, and other vast activities. Gov-ernments are therefore well suited to contribute to global-scale environmental action. Nevertheless, since many governments are designed to reflect the will of the people, for environmental responses to be undertaken with the necessary vigor, the people served by the governments must be ideologically supportive, at least in principle, of the gov-ernments' environmental goals.

Economics

Economic factors often contribute to a lack of effort toward environmental themes. If people do not have enough food or their families are in danger of running out of es-sential resources, they may be less willing to participate in environmental efforts. In fact, many individuals whose long-term livelihoods are based on natural resources, such as commercial fishers, will nonetheless exploit the resources that they rely on beyond the limits of sustainability, thereby jeopardizing their futures (National Public Radio, 2008). Societies in which there is a scarcity of essential resources such as food and water may have a particularly difficult time turning their attention to long-term environmental themes unless they can be directly coupled with short-term improve-ments in quality of life. In economic terms, these issues have a high discount rate. Some activists and philanthropists have suggested that the best work that can be done for environmental ends involves raising people's quality of life, and especially focusing on women's education, which correlates with reduced birthrate (Akmam, 2002).

In order for sustainability to succeed on a global scale, it needs to go hand in hand with an improvement in the quality of life of the world's peoples. It is unreasonable to expect a person who is starving to death to take action regarding the world's environmental issues. Luckily, there are numerous paths toward both sustainability and an increased quality of life. Many of them are paved with technological innovations.

Religion

Religion also factors into people's caring about environmental concerns (Gottlieb, 1996). The relationships that different religions create between humans and other living things profoundly affect how proponents of those religions act toward the world around them. Many religions paint the natural world as existing to serve humanity—for instance, "Do you not see that Allah has made what is in the heavens and what is in the earth subservient to you, and made complete to you His favors outwardly and inwardly?" (Quran 31:20, n.d.). Humanity is sometimes cast as the caretaker of the world. In the wake of new information regarding the world's environmental challenges, for example, many nondenominational Christian churches are embracing environmental issues, taking action in their appointed roles as caretakers.

Finding a Common Currency

Looking at the broad questions of world cultures and global sustainability, a UNESCO (UNESCO & UNEP, 2002) panel offered the following analysis:

The perceived separation between biological diversity and cultural diversity obscures the reality that both diversities are mutually reinforcing and mutually dependent. We cannot understand and conserve the natural environment unless we understand the human cultures that shape it. Each culture possesses its own sets of representations, knowledge and practices. Human action with respect to the environment, including management itself, is a social act and an expression of culture. (p. 8)

This connection between cultural and biological diversity and sustainability may be of help in striking a balance between the need for global environmental standards and the practicalities of local environmental action.

One problematic aspect of the quest for sustainability is the difficulty of comparing the impacts of different types of actions. Many different actions—such as recycling, flying in airplanes, eating beef, or using the stairs at work—have substantial environmental impacts. Each of these actions, and the thousands of other choices that we make unconsciously every day, have ramifications, but the details of those effects are rarely clear. Coming to an understanding of how these impacts relate to each other—which are the great offenders or great success stories—could help us adjust our resource allocation decisions in a more informed way.

The concept of a carbon footprint has developed into a common currency for many environmental concerns. There are numerous carbon footprint calculators on the Web, and carbon cap-and-trade systems have been proposed as ways to reduce civilizations' environmental impact. All of these systems rely on effective measuring of CO_2 emissions, which is nontrivially complex. Many other environmental concerns, though, may not be convertible into CO_2, such as habitat destruction and species extinction. Just as there are nontrivial challenges to establishing the carbon footprint of most objects and actions, similar challenges are likely to exist with any metric by which to compare environmental impacts due to the inherent complexity of the systems involved.

Individuals

In addition to the range of biological and cultural predispositions that affect human decision making, individuals develop concern for environmental issues through a range of processes, such as direct experience, learning, and social support.

Direct Experience
We humans place value on the world around us partly through our direct interactions with different ecosystems and environments. As an example, in some parts of the world a large percentage of people's food comes from foraging in undeveloped wilderness. In other places people go camping or kayaking for entertainment. Both of these cases involve people finding personal value in ecosystems where too much human competition detracts from that value. Based on these experiences, people may be predisposed to acting to ensure future access to these ecosystems.

Taking a story from my own experience, I have been a surfer for about eighteen years. In the past several years, however, since moving to California after graduate school, I have discovered firsthand that there is a correlation between rain falling and me getting a sinus infection. Apparently the rain washes pollution from the Los Angeles basin into the ocean, which then causes high bacteria levels, which then cause sinus infections and other illnesses. A bit of research revealed this correlation to be a well-known causative relationship (Connelly, 2008; Lambertson, 2009; Surfrider.org, 2009). Once it became apparent that my personal discomfort (sinus infections) was directly tied to an environmental problem (water pollution), I vowed to use each hour during which rain prevented me from surfing to work on environmental projects.

Learning

People also learn to care about environmental issues through various educational and learning experiences. The "Roper Report" suggests that 95 percent of Americans support the practice of teaching schoolchildren about environmental concerns (National Environmental Education Foundation, 2000). As will be discussed in greater depth in chapter 6, environmental topics are included in both the U.S. National Science Education Standards (National Academy of Sciences, 1995) and various state science content standards such as those in place in California (California State Board of Education, 2004). In addition to environmental learning in formal educational contexts, many people engage in self-directed learning about environmental topics, reading books, watching nature documentaries, and visiting parks to broaden their understanding.

Social Support

Various kinds of social support may also encourage environmental thinking. People may find satisfaction in participating in events organized by the Sierra Club or similar organizations. They enjoy receiving acknowledgment from their social group. Having other people to talk, celebrate, and commiserate with is an important part of human social experience. Moreover, having other people with similar lives going through the same environmental transformations could be helpful to the process of sticking with new habits and decisions.

People may have an easier time adhering to some behavioral choice if they believe that their actions are being mirrored by many other people, and thus are having a significant impact collectively. Besides the feeling that other people are acting in the same way, there is an awareness that, if many people are taking a certain set of actions, then support structures for those actions will spring up, if they do not already exist. For example, the more people buy electric cars, the more financial incentive there is for someone to open a charging station nearby. As will be discussed in chapter 8, many different systems are arising to support collective action.

Challenges

Despite various personal and social factors that encourage environmental thinking and action, there are a number of challenges that stand in the way of sustainable living on a large scale. Several cases come to mind from my own personal experience, each of which (temporarily) dampened my environmental enthusiasm.

One night when I was working late at the research lab at MIT where I did my graduate study, I saw the cleaning crew come around with their large trash can. As they went through each room, they emptied both the small trash cans and the recycling bins into the same large trash can. Seeing this made me realize how many stages there are between many environmental actions and their ultimate impact on the world. Action or inaction by other individuals at each of these stages can render irrelevant any actions that one has taken. In the cleaning crew's defense, I have seen many white paper recycling bins filled with plastic bags, Styrofoam containers, apple cores, and an assortment of nonwhite paper objects. Nevertheless, having the recycling emptied into the trash undermines some effort and a lot of environmental goodwill. This particular event really drove home for me how important it is for organizations, as well as individuals, to take action and ensure vigilance.

A second example came when my old car broke down, and I was contemplating buying a hybrid Toyota Prius, which had received a great deal of good press for being the environmentally friendly choice. My friend then offered that the Ni-MH batteries in the Prius are bad environmentally—so much so that the environmental footprint of the Prius was similar to an SUV. He later offered that he could not back that statement up with evidence, but on researching the Prius later it does appear that the manufacturing of the car off-loads a lot of the environmental impact of it from visible issues such as gas consumption onto less obvious ones such as battery recycling. An awareness of the entire life cycle of an object is critical to understanding its impact.

A third example comes from a course that I teach, titled Social Analysis of Computerization. I cover a number of issues in this class relating to computing and the environment. When I teach the environmental section, I ask students why they think that environmental issues are important. Once, I had a student raise his hand and offer, "There's nothing we can do about them [environmental issues], so they don't matter." I was a little taken aback by this stance. We discussed his proposition a bit, so that I could understand it fully. I came to recognize that this student and presumably many others felt that environmental issues were too vast for them to have any impact. In addition, I felt that there was some resistance to this set of ideas—the student was uncomfortable with the whole topic. I did my best to present concrete illustrations of ways that we all could do something about environmental issues. I also offered that, even if there were nothing we could do about it, a large population reduction would, in fact, matter a great deal. The student agreed, if somewhat grudgingly.

These examples are unique to my own life, but are indicative of the kinds of experiences that might discourage a person from engagement with the effort to live sustainably. All three arise from the overwhelming complexity that underlies most environmental issues.

Broadening Horizons

The key problem faced by many efforts to address environmental issues is one of horizons. Essentially, we humans do not tend to concern ourselves with issues that transpire on scales of time, space, or complexity beyond certain limits; they are beyond our horizons. To demonstrate the difference between human and environmental time horizons, for example, consider the following. The interest rate for a thirty-year mortgage is currently 5.10 percent per year. This rate provides some insight into the time horizon with which many people currently think, at least those living secure enough lifestyles to worry about mortgages. (Human time horizons change profoundly when the individual is in danger, faced with starvation, etc.)

For environmental issues, the time horizon could be modeled by the current species extinction rate—approximately 0.25 percent annually, based on approximately twenty million existing species, and fifty thousand going extinct each year (Olson, 2005). Seen on the time scale of an individual person (less than one hundred years), the species extinction rate does not seem to have much effect. A hundred years later, the person will likely be dead, inflation may have caused the value of uninvested money to have dropped dramatically, and yet 80 percent of the species will still be alive. Looking at these same phenomena on a longer time scale, however, it becomes apparent that species are declining at an alarming rate. There are nonlinearities in environmental impacts as well—thresholds where the graph is not smooth (e.g., ecological tipping points where the loss of one additional species makes the loss of many others inevitable). Nevertheless, getting people to care about something that will happen long after they have died remains a challenge.

To address environmental issues effectively, human civilizations must adopt horizons broad enough for us to understand problems that occur over hundreds of years and thousands of miles, and take meaningful action based on that knowledge.

Despite our propensity for satisfying short-term needs, humans are nonetheless able to postpone the gratification of our immediate desires for long-term ends. Instead of grabbing someone else's food from their table at a restaurant, we order our own food and pay for it at the end. Even better, we are sometimes able to develop ways of gratifying our immediate desires while serving our long-term goals. Intelligence, one of the areas in which humans excel, can lead to better short-term *and* long-term outcomes, for instance, by discovering opportunities for efficiency that both increase profitability and reduce resource consumption (von Weizsäcker, Lovins, & Lovins, 1998).

One of the great human abilities is our ability to see into the future. In addition to our ability to eat when we are hungry, we are (sometimes) able to save money, not eat too many fatty foods, and spend multiple years in educational efforts to attain future benefit. This ability to sacrifice immediate benefit for even greater future benefit—postponing gratification until a later time—is a key part of the reason humans have thrived around the world.

Still, we are only willing to perform certain kinds of self-sacrifice for future gain. If I were to ask you if you would like five dollars now or six dollars fifty years hence, I expect that most of the readers would take the money now. Why is that? It might be because you would think you could make more than 20 percent gain over the next fifty years. It might be because you suspect that I might not still be around to give you the six dollars in fifty years. It might be because you had heard the old saying, "A bird in the hand is worth two in the bush." The economic concept of *present discounted value* is based on the notion that money now is worth more than money later.

People are most effective when thinking on the scale of a single human existence, rather than on those that are much longer or shorter, or bigger or smaller. This predilection lies at the heart of the world's environmental issues. It is hard for us to understand issues at the scale of the global ecosystem, and the scale of several decades or centuries. While graphs or news stories may help to make the issues clearer, we still are able to put these long-term and large-scale issues out of our minds in favor of shorter-term, smaller-scale issues such as, "Do my children have enough food this morning?" or "Did I remember to turn off the stove?" This focus on narrow problems is not unique to individuals; corporations and governments are often shortsighted as well. According to the director of an environmental NGO who was cited in the *Smart 2020* report, "Even though there is a good business case, Chinese companies either do not trust the business case or take a short-term view. They prefer upfront cash to promised future cash flows" (Climate Group, 2008, p. 35). When people do think about phenomena that are psychologically distant, though, they tend to think about those issues in high-level and abstract ways that may lend themselves to addressing environmental issues (Trope, Liberman, & Wakslak, 2007). IT can help people engage with these distant phenomena and provide the tools to take local action.

Although many people adhere to narrow horizons when planning for the future, there is a compelling argument that the lack of concern for global climatic disruption's impact only makes sense in the context of positive rates of return. When the analysis includes the possibility of declining worldwide consumption, even if this downturn does not start for fifty years, making investments to mitigate climatic disruption is economically rational (Dasgupta, 2006).

The fact that most of us are more focused on our own local problems does not solve the broader environmental problems looming on the horizon. Extending the spatial metaphor, it is critical that we attend to these issues in the distance before they become insurmountable obstacles right in front of us. Luckily, humans have a long history of using technology to augment our capabilities to address large problems such as hunger and warfare. Technology may help address the growing global environmental concerns as well.

People are good at thinking about events that occur on human scales of time and space. We evolved to deal with problems of food gathering, immediate danger avoidance, and proximate reproductive success. Since we did not have the capabilities to understand or act on phenomena that occurred over centuries or thousands of miles, we did not develop the cognitive apparatus to comprehend these scales effectively. As a result of scientific advances, however, we are now in a unique position in the history of living things to work together as a culture to understand problems that occur on much longer time scales and over much larger areas, and enact well-informed change to make life better for ourselves and our descendants. Nonetheless, we need theoretical and technological assistance to help us think about these phenomena and act effectively. We need tools and techniques that will enable us to comprehend vast (from a human point of view) phenomena. Understanding the commonalities across different kinds of change on different scales of time and space, from near-term disasters to vast climate change, can enable people and institutions to launch timely, efficient, and appropriate responses in ways that previously were not possible.

A human-centered approach to the world's environmental issues can help explain how humans and our IT have contributed to the current environmental challenges, and how humans and IT can help resolve these issues. This chapter has discussed key factors in humans' perspective on environmental themes. In particular, it has focused on the biological underpinnings for the narrow horizons that dictate much of human behavior, and sought to understand how cultural phenomena enable us to operate on broader scales.

The next chapter considers how technologies and the social systems that interact with them help address this issue.

4   The Role of Technology

Technology is neither good nor bad; nor is it neutral.
—Melvin Kranzberg, 1986, p. 545

Many different definitions of technology are available, each of which provides a different view into the role technology plays in human affairs. The Greek roots—*techne* + *logia*—suggest that it is the "study of craft." The *Oxford English Dictionary* provides the following definition: "A discourse or treatise on an art or arts; the scientific study of the practical or industrial arts" (OED.com, 1989b). Computing pioneer Alan Kay is reputed to have stated that technology is "anything that wasn't around when you were born."Author Arthur C. Clarke offers that "any sufficiently advanced technology is indistinguishable from magic." (Clarke, 1973, p. 21) (Author Larry Niven is supposed to have inverted this comparison, suggesting that "any sufficiently advanced magic is indistinguishable from technology.") Taken together, these descriptions provide some functional, temporal, and practical sense of how technology impacts humanity.

Tool use is closely bound up with technology. Humans are particularly crafty with our use of tools. We use tools to affect our world. In fact, in an earlier era tool use was seen as a characteristic that distinguished us from "dumb animals" until those nearly-as-crafty chimpanzees demonstrated that they could use tools. We use tools to make other tools, which nearly separated us from the chimpanzees, until researchers found them doing it, too. We are, however, perhaps the only organism to have a firm cognitive grasp of the recursive nature of tool use; we use tools to make tools to make tools to make tools, and understand that we can keep layering on more tools until we get what we want or run out of resources. Researchers have suggested that chimpanzees have a working memory of size two (that is, that they can hold only two items in their memory at a time), and therefore are not capable of recursion (Read, 2006).

Because humans are so skilled with the development and deployment of new tools and techniques, technology often ushers in profound changes to the social status quo. These changes are not always welcomed by all parts of a society. Niccolò Machiavelli

(1532) offered that "the innovator has for enemies all those who have done well under the old conditions, and lukewarm defenders in those who may do well under the new." We have seen vast changes in global civilizations over the last decade, century, and millennia, from the development of the personal computer, to the utilization of fossil fuels, to the widespread use of the books enabled by the printing press.

IT is among the most powerful and rapidly growing forms of technology in use around the world. It has enabled new forms of communication, new ways to store and manipulate data, and new kinds of entertainment, among its many impacts. The social contexts in which IT systems operate are critical to understanding how future IT systems could and should be designed. The interactions between technologies and the social systems in which they are embedded are of profound importance to how technologies affect people. This chapter explores some of the ways in which technology and sociotechnical systems factor into the world's current environmental challenges.

The [information and communications technologies (ICT)] sector has a unique ability to make energy consumption and GHG emissions visible through its products and services. Radical transformation of infrastructure is possible only if it is known where inefficiency occurs throughout the processes and workflows of various sectors in the economy. ICT can provide the data, which can be used to change behaviours, processes, capabilities and systems. Although isolated efficiency gains do have an impact, ultimately it will be a platform—or a set of technologies—working coherently together that will have the greatest impact. (Climate Group, 2008, p. 14)

Beyond just providing the data for human decision making, IT can change the ways in which people relate to environmental questions, leading to new ways of thinking, living, and engaging with the world around us.

Technology as Multiplier

Technologies are designed to provide a multiplier effect on human effort. They empower us to take action on larger scales of time and space with less personal effort. Eyeglasses help us see more clearly; tractors help us till fields; bombs help wage war; and computers help process information. By magnifying the impact that individuals and groups have on the world, technology is an enabler of human behavior.

The fact that people use technology to do more with less effort has had some peculiar social effects. Clay Shirky discusses the impact of technology on society, starting with the Industrial Revolution. Citing an unnamed British historian, he suggests that gin was the most important technology in the early stages of the Industrial Revolution. Faced with the social transformations involved in moving from rural areas to cities and greater amounts of free time, people drank alcohol rather than engaging in more constructive endeavors. Shirky (2008) sees a parallel in the twentieth century:

If I had to pick the critical technology for the 20th century, the bit of social lubricant without which the wheels would've come off the whole enterprise, I'd say it was the sitcom. Starting with the Second World War a whole series of things happened—rising GDP per capita, rising educational attainment, rising life expectancy and, critically, a rising number of people who were working five-day work weeks. For the first time, society forced onto an enormous number of its citizens the requirement to manage something they had never had to manage before—free time.

And what did we do with that free time? Well, mostly we spent it watching TV.

Shirky goes on to talk about the positive impacts that could come from human civilizations using some of the "cognitive surplus" that people currently pour into television watching for more constructive ends. He provides a new unit of measure—*the wikipedia*—to help understand the magnitude of the surplus. A wikipedia represents one hundred million hours of human thought—the amount of time used to create all parts of Wikipedia, from the computer code to the content. While this number may seem large, Shirky counters that people in the United States alone watch two hundred billion hours of television every year, or two thousand wikipedias. Perhaps more dramatically, U.S. television watchers spend a full wikipedia of thought every weekend just watching the advertisements (Shirky, 2008).

This idea that people could take some of the billions of hours spent watching television and instead channel it into projects that involve working collaboratively with others to create, innovate, or construct knowledge lies at the heart of what Henry Jenkins (2006) has called "participatory culture." Along similar lines, John Thackara (2005) suggests that we should "treat 'content' as something we do, not something we are sold" (p. 8). Because of the multiplier effect of technology, we can have a vastly greater impact with our time and effort; how we use this augmented effectiveness will be an important factor in the coming years. If even a fraction of the people in the world used a greater part of their time developing Green IT systems, the collective impact could be very great.

A Double-edged Sword

The cultural mythologies surrounding technology can take on radically different forms. One such mythology is that technology can save us from ourselves, positing that new technological innovations will banish our current ills and enable a techno-utopia. Another is that technology will be humanity's demise, as featured prominently in movies such as *The Terminator* and *The Matrix*, where people's technology becomes sentient, gets uppity, or otherwise runs amok. Technology often feels strongly valenced (i.e., polarized along a good-bad axis).

Rather than technology itself being inherently positively or negatively valenced, though, it is only as good or bad as the uses to which it is put by people. Technology empowers humanity; whether technology is good or bad therefore depends on how one views the human activities empowered by it.

This is not to say that specific technologies do not provide a bias on the aspects of human activity that it empowers. This distinction between the broad topic of "technology" (the study of practical arts) and "a technology" (a specific tool or technique used for some purpose) is important in this analysis. As will be discussed later, specific technologies are designed, and as such tend to be particularly good at some specific task. The specifics of each technology and the ways in which people are able to interact with it bias which human behaviors are empowered by it.

Technology as a whole is only as good or bad as humanity itself. For optimists, technology might seem like cobblestones on a path to utopia; it helps us succeed in being good, just a little faster than we did before. For pessimists, technology helps us destroy things and hurt each other like we always have, but now a little bit faster and more effectively than before. Instead of looking to technology to solve our problems or waiting for technology to destroy us, we instead need to look to people to solve problems (facilitated by technology) or figure out how to keep people from destroying each other (with the job made easier by technology). Humans create technology, humans take action, and humans use technology to make our actions have more impact. What impact we will have depends on ourselves more so than it does on technology as a whole.

Underlying a belief in the net usefulness of technology is a basic optimism about humanity—that the technologies that will be designed, and the uses to which people will put them, will make things better in the future. If humans are seen as inherently bad, then technology will only make us destroy ourselves more rapidly. This book tends to the more optimistic approach toward humanity and thus humanity's use of technology.

Technologies Are Designed

While it might be more soothing to think about technology as being somehow responsible for humanity's actions, thus obviating the need for our own accountability, it is important to remember that technologies are designed by people, either explicitly during some manufacturing process or implicitly during their recruitment for use. The way the technology itself works makes it better for certain tasks than others. Hammers are good for hammering, and less good for removing screws or making omelets.

The process of creating a technology allows designers to have some control over how that technology will be used (Norman, 1988). By manipulating many different aspects of the technology—such as functionality, interface, documentation, and marketing—designers can affect how people will tend to use it.

For example, in the design of a hammer, the placement of a soft grip that is contoured to accommodate fingers may make it more likely that people will hold the correct end of the hammer for its intended usage. One could certainly hammer, albeit less effectively, by holding the curiously shaped metal end and whacking a nail with the softer, lighter end. On experimentation with a new tool (which people are reasonably skilled at), however, people may tend to gravitate toward the usage style intended by the designers.

The design process is inevitably constrained by the past experiences and worldviews of the people doing the designing. Ultimately, technologies are designed by individual humans or teams of humans, rather than by corporations, nonhuman animals (except perhaps for the occasional tool-using chimpanzee), or ecosystems. Therefore, technologies tend to be designed for individuals or teams of humans, rather than for the needs of other entities. While human designers sometimes endeavor to enable their designs to work well for corporations or serve the needs of animals, we do not have as intuitive a grasp of the constraints and opportunities of systems designed for use other than by individual people; we are thus less skilled at designing technological systems that serve humanity in general or all living things as a whole.

People Use Technologies, But Not Always as Intended

To understand how technologies impact the world, it is necessary to consider not just the design of the technologies but also the ways in which people use them. Human systems are changed by the technologies that they use; examining the wide range of interactions between humans and technological systems can shed light on the tight entanglement between people and technology.

Same Goal, Different Technologies

When undertaking a given task, people may have a variety of technologies at their disposal. When trying to measure a window, I could use a tape measure, a piece of string, and a ruler, my hand (which I know is about 9 inches from the tip of my thumb to the tip of my pinkie), or a piece of 8.5 x 11 inch paper (folded in half or quarters if I need to increase the accuracy of the measuring). Different technologies offer different advantages and challenges in undertaking a certain task—ease of use, accuracy, availability, social acceptability, and so on.

There are numerous decisions involved in selecting the best technology for a given task. For example, should one use a technology that is at hand, go find a superior technology that exists but is not immediately available, or invent some new technology to address the problem? Understanding the relative merits of many different technologies for a given set of tasks is a large part of what makes people "experts" in certain domains.

The merits of a given technology may be subjective, based on what else is available and practical to use. As my classmate at MIT Ken Russell once said while discussing the variety of debugging tools, "When printf's all you have, it's still pretty good." The C/C++ function "printf" simply prints text to the output screen and may be inferior to more newfangled debugging systems, but if those systems are not installed or otherwise available, printf may still be perfectly serviceable. A corollary of this idea is that the suite of technologies at a person's disposal changes the way any new technology is perceived. Providing home-builders with hammers when they have only rocks for tools may be a big improvement, but if they already have nail guns, the hammers may not have the same luster.

Same Technology, Different Goals

Also relevant are the ways in which people use the same technology to pursue different goals. People will co-opt technologies for a variety of different goals, sometimes effectively, and sometimes with disastrous results (such as in the story passed among technical support staff and systems administrators of a novice computer user who attempts to use the CD drive on a computer as a cup holder).

As the old saying goes, "When you have a hammer in your hand, everything looks like a nail." People with experience using a certain tool or technique may apply it in a wide variety of settings, only some of which may be well matched for it. This multipurposing of technologies may be intended or unintended by the original designer. A computer may be used to write a paper, play a game, or send an email—all uses intended by the designers—but it may also be used to hammer a nail into a piece of wood—a use presumably not intended, since a computer makes an expensive, fragile, and not particularly effective hammer.

As mentioned in chapter 1, Simon (1996) wrote: "Everyone designs who devises courses of action aimed at changing existing situations into preferred ones" (p. 111); similarly, everyone designs who co-opts technologies for new purposes. In his book *The Hidden Connections*, Fritjof Capra (2002) discusses the interplay between emergence and design. He offers a challenge: "to find the right balance between the creativity of emergence and the stability of design" (p. 121). While technology is designed, the unexpected phenomena that emerge around new technologies demonstrate that designers can only be partly in control. By leaving room for new uses to emerge, designers can help users of a technology to be designers as well, developing new ways to utilize existing technologies.

Because people are willing and able to repurpose technologies, understanding the human context of a technology is necessary to understand how the technology will be used (rather than how it *should* or *could* be used). That context is multifaceted, involving many different aspects of human lifestyles and desires. Understanding social context has been growing in importance in human-computer interaction and other fields over the last several decades (see, for example, Dourish, 2001; Suchman, 1987).

As we rush along, embedded in our physical and social contexts, in pursuit of what we want, designing and innovating as we go, we often bring about consequences beyond those we initially intended. Sometimes these side effects are good, such as when an innovation called the Internet, intended to facilitate research and maintain the military chain of command during a nuclear war, allows me to have movies streamed to my computer. At other times they are less good, such as when the chemical technology dichlorodiphenyltrichloroethane (DDT), intended to increase agricultural productivity, accidentally poisons large portions of the global ecosystem (Carson, 1962). To be fair, it is important to represent the wide array of impacts of any of these technological advances: streaming movies uses electricity and keeps me from going out of the house, while DDT helped reduce malaria among soldiers in World War II. Regardless of the range of impacts, it is not possible for innovators to comprehend all potential uses and effects of a technology.

Same Technology, Same Goal, Different Usage

To illustrate another way in which human ingenuity affects the use of technologies, consider the abundance of different ways to use the same technology in the apocryphal story of a certain physics student. When asked how to measure the height of a building using a barometer, the student offered seven different ways to do so—tie a piece of string to it, lower it to the ground, and measure the string; drop it and measure how long it takes to hit the ground; set it next to the building, measure the length of the shadows of both building and barometer, and use proportional arithmetic; swing the barometer like a pendulum both at ground level and the top of the building, finding the height via the gravitational restoring force; walk up the emergency exit and measure the building in barometer lengths; measure the air pressure at the top and on the ground (presumably the desired answer); and go talk to the janitor and say, "If you would like a nice new barometer, I will give you this one if you tell me the height of this skyscraper" (snopes.com, 2007).

We humans are remarkable in our ability to come up with an abundance of viable solutions even when our actions are constrained. This very robustness in our capacity to adopt and co-opt new technologies makes the study of technology and its interactions with society challenging as well as rich with opportunities. Many of these opportunities will have significant impacts on the environmental sustainability of human civilizations.

Metaphors

A large part of how people use technological systems is determined by the metaphoric framing through which we engage with those systems. Metaphors are a crucial part of how people think, helping us construct our understanding of the world we inhabit (Lakoff & Johnson, 1980). We use metaphors to understand new objects, processes, ideas, and settings by relating them to phenomena that are already familiar.

The history of computing is full of many different metaphors. Computers show a clear legacy of the "computer is a typewriter" metaphor, featuring the printing of text documents, alphanumeric keyboards, and especially the return key (which is short for a typewriter's "carriage return"). Notebook computers have a metaphor right in their names—a connection that elicits note taking, drawing, and portability.

Among the most pervasive metaphors currently in use in computing is that of the virtual desktop. A number of alternatives to the desktop interaction paradigm have been proposed, including an agent-based theatrical one with a speech interface (Negroponte, 1991) and a 3-D windowing metaphor where the computer is a collection of portals through which a user passes to access different parts of a 3-D world (Smith, Kay, Raab, & Reed, 2005). These metaphors cause the user to approach the computer with a predominantly visual and aural interaction.

A conscious attention to the metaphors relating to various technologies can enable the design and deployment processes to help a technology have the intended effect. While the embedding of metaphors in technological innovations is an inevitable effect of the process of human ingenuity, doing so only via intuition, rather than also reflecting consciously on the metaphors in use, may limit the design process.

Environmental Impacts of IT

While the main focus of this book involves the potential benefits of Green IT, it is important to consider the environmental impact of the IT sector itself. A Gartner presentation titled "Green IT: The New Industry Shockwave" (Mingay, 2007) organizes the environmental impacts of IT systems into three categories: first-order, second-order, and third-order effects. The first-order effects are those that arise as a direct result of the existence of the technology. Among the first-order effects, Gartner includes GHG emissions, e-waste, hazardous substances, and the use of scarce and nonrenewable resources. While the second- and third-order effects (e.g., the optimization of transportation systems) are discussed throughout the rest of the book, this section addresses explicitly the first-order environmental challenges that accompany IT.

GHG Emissions

One of the key challenges that IT faces is the issue of power consumption and its accompanying GHG emissions. Due to the power consumption of all the processing that humans use, from Web searches to mobile phone calls, one's computational footprint is becoming a nontrivial part of one's carbon footprint. There was a recent debate online over the carbon footprint of a search of the World Wide Web via the Google search engine (Kincaid, 2009), with Jonathan Leake and Richard Woods (2009) suggesting that a search produces 7 grams of CO_2 (about half the amount needed to heat a pot of tea), and Google rebutting that figure with its own estimate of 0.2 grams of CO_2 (Hölzle, 2009). Regardless of the precise figure, power consumption is clearly a major issue for the search company; Google has begun putting its server farms on the banks of the Columbia River in Oregon, in its quest to find economical power supplies for the vast amount of processing that it performs (Strand, 2008).

As another example, Nicholas Carr (2006) did the math to demonstrate that an avatar in the online world "Second Life" consumes approximately as much energy as a real-world Brazilian citizen. When I asked the chief technology officer of Linden Lab, maker of Second Life, about this article, he offered various ways in which the calculations were off by several multiples. Nevertheless, even if a Second Life avatar has the impact of just one quarter of a Brazilian, it is still striking evidence of the substantial power draw of IT.

Overall, the ICT sector is responsible for approximately 2 to 2.5 percent of global CO_2 emissions (Mingay, 2007; ITU, 2008). Figure 4.1 shows how this amount is broken down across various areas.

The power consumption issues of IT are even more of a concern when one considers the great capacity that each human has for being supported by computing. While one individual can only wear as much clothing as will fit on one body and carry only so many physical objects, a simple mobile phone could be a portal into a vast amount of computing resources. As search, multimedia, and artificial intelligence techniques become more powerful, they will have correspondingly more power demands. Despite Moore's Law, which predicts that we can fit more and more processing into a smaller and smaller space, we are still increasing our IT power demands dramatically.

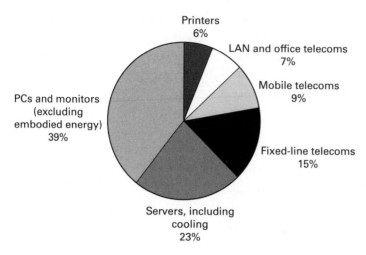

Figure 4.1
ICT's global carbon footprint. From data in (Mingay, 2007).

Electronic Waste

A great deal of effort in technology corporations and computing researchers alike is the generation of the "next big thing." However, much less attention is paid to what happens to the last big thing. Microsoft's Vista operating system, for example, did not run on many of the computers in use around the world, thereby necessitating the replacement and discarding of the old computers. Discarded technology is a substantial environmental issue.

The rapid turnover rate of IT and resulting obsolescence of old computing devices is leading to a great deal of e-waste around the world. This waste is often discarded in countries other than those in which it was produced and used. This transportation shifts much of the e-waste burden from the industrialized nations, where technology is more ubiquitous, to the nonindustrialized nations, where regulation and oversight is less vigorous.

Basel Action Network is among the most well-known watchdog organizations regarding e-waste. Its videos, such as *The Digital Dump: Exporting Re-use and Abuse to Africa* (Basel Action Network, 2008), showcase the problems of e-waste in the locations where it is ultimately disposed (see figure 4.2).

Hazardous Substances

E-waste contains a variety of toxic chemicals, such as the lead used in CRT screens. The effect of toxic residues that enter water systems and food chains from discarded

Figure 4.2
Migrant child from Hunan Province sits atop one of countless piles of unrecyclable computer waste imported from around the world. Guiyu, China. December 2001. Photo and caption © Basel Action Network. Reprinted with permission.

computers and other technologies is rapidly becoming a global problem. Although manufacturing practices for appliances and digital technologies are improving, using fewer toxic materials and greener production processes, the sheer scale of production in a global economy results in a growing e-waste problem. Toxins from discarded technologies end up in the water and food of the poorest countries where waste disposal is an industry, and where disposal uses expedient rather than safe techniques (see figure 4.3). Consumer programs to recycle and reuse technology are in place in many locations around the world, but through a lack of understanding, people do not avail themselves of these opportunities.

Scarce and Nonrenewable Resources

The production of IT devices uses an array of scarce, nonrenewable, or otherwise problematic resources such as lead, mercury, and cadmium. Nokia (2009) provides a list of the twenty-six categories of substances that have been banned or are otherwise re-

Figure 4.3
Molten television. Use of brominated flame retardants and heavy metals in plastics by manufac-
turers such as Sony create dangerous emissions when burned and disposed in this way. Lagos,
Nigeria. Photo and caption © Basel Action Network. Reprinted with permission.

stricted in the production of its technologies (Nokia, 2009); nevertheless, some or all of
these materials may still be in use by other technology manufacturers.

Coltan, in particular, has become indicative of the global problems caused by the
materials used in IT systems. This metallic ore is refined to produce tantalum, which is
used in the production of capacitors, which are found in nearly all mobile phones, laptop computers, and many other devices (Democratic Republic of Congo, 2009). Coltan
mining is implicated in the ongoing civil war in the Congo and has contributed to the
endangerment of the eastern lowland gorilla (National Public Radio, 2001). Nevertheless, the ongoing demand for coltan makes it unlikely that mining it in the Congo will
cease any time soon.

The IT sector is growing rapidly. The *Smart 2020* report forecasts that the global
number of personal computers will increase to over four billion by 2020, more than six
times as many as there were in 2002 (Climate Group, 2008), and many other kinds of

devices are spreading rapidly as well. The use of scarce resources and potentially dangerous materials is likely to continue to be a substantial problem for the IT industry and the ecosystems that are impacted by it.

IPAT

Technology has long been responsible for helping people exploit the world around us as effectively as possible. The multiplier effect of technology as it relates to environmental impact is featured clearly in an equation called IPAT. This equation posits that environmental impact (I) equals population (P) times affluence (A) times technology (T) (Commoner, 1972). Essentially, the environmental impact of a group of people is a product of the size of that group with the amount of consumption they engage in and the effect of the processes they use to engage in that consumption. While this equation has some flaws (e.g., twice population does not lead to precisely twice the impact), it nevertheless highlights the key role that technology plays in the impact of humans on our environments.

The Downside of Techno-Utopianism

A more subtle impact of IT that makes it potentially problematic for environmental action lies in the broad appeal of techno-utopianism. For many people, there is a sense that technology has fixed our problems in the past, and so it is really just a matter of time until the technologists come up with a whiz-bang gadget that will solve all our future problems as well. Unfortunately, this hope enables an abdication of personal and collective responsibility for the world's unsustainability. People will need to use IT systems in sustainable ways; designers can only provide us with systems that lend themselves to sustainable use. While one of the key points of this book is that IT may, in fact, be able to assist with a concerted human response to the growing environmental concerns, it is important for us to remember that it is ultimately a *human* response; technology can simply aid in the effort.

Making IT Green

As discussed earlier, technology enables human activities, and as such is a double-edged sword, particularly with regard to environmental impacts. The preceding section considered the problematic aspects of IT. According to the *Smart 2020* report, however, the benefits of IT could outweigh its harm:

ICT can make a major contribution to the global response to climate change. It could deliver up to a 15% reduction of [business as usual] emissions in 2020 (7.8 GtCO$_2$e), representing a value of

€553 billion ($872.3 billion) in energy and fuel saved and an additional €91 billion ($143.5 billion) in carbon saved assuming a cost of carbon of €20/tonne, for a total of €644 billion ($1,015 billion) savings. This saving in CO_2e is more than five times the size of the sector's own footprint and its size demonstrates the important role an advanced communications platform can play in the transition to a low carbon economy. (Climate Group, 2008, p. 51)

The remainder of this chapter focuses on how technology in general and IT in particular can help human civilizations become more sustainable, and how the technologies themselves can be made more sustainable as well. The broad challenge is to build and use technology such that it enables us to live with an acceptable quality of life without consuming more total natural resources than the nonhuman world is capable of replenishing.

The central problem in our understanding of the need for sustainability is our relatively narrow horizons for thinking about the future, distant places, and great complexity. Since technology is often built to allow us to achieve our goals more effectively, these narrow horizons have become built into our technology. These horizons become self-perpetuating; the implication of our disposable technology culture is that we have a high rate of discount for how good something will be later, and that new technologies will be in place to supplant it when that time comes.

Design for Sustainability
Several authors have substantially influenced my thinking about role of sustainability in the design process.

Blevis's (2007) sustainable interaction design paper and his upcoming book with Elaine Huang (Blevis & Huang, forthcoming) provide an excellent description of the relationship between sustainability and interaction design. They offer a framework for understanding the material effects of interaction design and seeking to reduce the impact of interactive systems. Specifically, they supply a rubric for assessing the impact of designs, including consideration of ten items: disposal, salvage, recycling, remanufacturing for reuse, reuse as is, achieving longevity of use, sharing for maximal use, achieving heirloom status, finding wholesome alternatives to use, and active repair of misuse. The concept of an "heirloom technology" is especially noteworthy—a technological system that is not designed for prompt disposal, but rather one that gains value with age, use, and history. In addition, they offer several principles for enabling sustainability concerns to factor into the design process: linking invention and disposal, promoting renewal and reuse, promoting quality and equality, decoupling ownership and identity, and using natural models and reflection. By reimagining the role that technologies can play in society, these researchers are helping enable interaction designers to lessen the environmental impact of their designs.

In *Cradle to Cradle*, McDonough and Braungart (2002) reenvision the notion of design, rejecting the traditional "cradle to grave" premise, in which design contributes to a life cycle that transforms raw materials into waste while providing usefulness to people. Instead, they propose that the design process should facilitate the process of providing usefulness to people by mirroring natural ecological systems in which the end products of one process become the raw materials for another process. Rather than having manufactured goods end up in landfills, they should be designed in such a way that, at the end of their usefulness to one client, their elements can be readily repurposed, serving as the "technical nutrients" for other products and services. The waste from every product becomes the raw materials for some other product, integrating reuse and "upcycling" (the process of using waste materials to create new products). McDonough and Braungart offer many examples of how this way of thinking about design already occurs, and potentially could occur across a range of industries and domains.

In *Factor Four—Doubling Wealth, Halving Resource Use*, Ernst von Weizsäcker, Amory Lovins, and L. Hunter Lovins (1998) explore ways of making many different areas of human civilization more sustainable by looking for inefficiencies in resource use, and redesigning to make systems increase their value to manufacturers by decreasing their consumption of resources. By focusing on resource efficiency, designers can help develop new kinds of systems that are a significant advance beyond current ones in terms of both profitability and sustainability.

Similarly, in *In the Bubble*, Thackara (2005) presents ideas about how to use design to shift from products to services, enabling new ways of living that are more socially and environmentally responsible. He uses the concept of "lightness" to think about how systems should be designed. Citing *Natural Capitalism* (Hawken, Lovins, & Lovins, 2000), Thackara claims that the waste matter in the production of a laptop is four thousand times the weight of the laptop. He suggests that, instead of this kind of product, which exerts a vast (and often unseen) weight on the world, we need services that can help us have a similar quality of life without such a focus on the consumption of goods.

We need fewer technologies that become obsolete or wear out quickly, and more technologies that, by their form and function help reduce our discount rate and improve our ability to delay gratification. We also need fewer that play into the disposable, short time horizon culture that has led to the current environmental crisis.

Economists and politicians have to square the utter necessity of a rapid and controlled shutdown of emissions from fossil fuel burning with the human needs of civilization. Economic growth is as addictive to the body politic as is heroin to one of us; perhaps we have to keep the craving in check by using a safer substitute, an economist's methadone. I would suggest again that the mobile phone, the internet and entertainment from computers are moves in the right direction; they use time and energy that might otherwise be spent travelling by car or aircraft. Moreover, there is information technology and the efficient use of energy, for example using the ultra efficient white light emitting diodes (WLEDs) to see at night. Should technology of this kind become the main source of economic growth it would let us spend our lives harmlessly and fill some of the time that now we use in fuel-consuming travel. To an extent we are evolving that way. (Lovelock, 2006)

James Lovelock makes a good point: that technology can help us live our lives in ways that are less energy intensive than flying or driving. Nevertheless, mobile phones and other computational devices generate an abundance of e-waste and consume a large amount of power. As computational systems of many different kinds become faster and more capable, they are likely to become even more resource intensive.

Technologies and sociotechnical systems can help address the world's environmental concerns in several different ways, as will be discussed below; they can help us think longer term, they can help us live longer term, and they can be designed to last longer. In all three of these cases, they can supplant older technologies that are less effective in these ways (although there is certainly a concern that the supplanting of previous technologies is, in itself, playing into the rapid turnover rate of computing technology). While personal lifestyle changes are only one part of a viable solution, with political, corporate, and other institutional changes needing to take place as well (Dourish, 2008), technologies that assist us in changing our minds can ripple through these broader-scale domains of human activity.

Standards
One of the characteristics of IT that makes it particularly well suited to environmental issues is that it features global standards. Civilizations thrive because of viable standards, such as laws (standards for behavior), money (standards for exchange), or calendars (standards of time). IT provides another such standard, with computers around the world agreeing about TCP/IP and other formats for exchanging digital information.

One of the reasons for the ICT sector's success is that it has developed layers of internationally standardised ways for machines to communicate with one another. International dialling codes, which have been around for more than a century, or the .com domain names are both obvious standards that allow rapid innovation and rollout of services. Protocols, or the rules that allow machines to send messages between each other, are hidden to the average user but underpin the internet's rapid development. TCP/IP actually refers to a set of interconnected protocols that support email and internet connectivity. XML, one of the specifications that underpin blogging or social networking applications, also allows the development of applications that manage an organisation's supply chains. A stack of interoperable protocols allowing for the communications between devices, applications and the standardisation of information exchange would allow more effective monitoring, control and minimisation of energy use and carbon emissions. (Climate Group, 2008, p. 55)

Standards are necessary for efficient action on a large scale; otherwise, a great deal of time and energy is lost negotiating about formats and procedures, rather than addressing the issue at hand. IT standards enable communication across the world, with a broader reach than individual languages, currencies, and even calendars (e.g., in some parts of the world 1/12/09 means January 12 and in others, December 1). Humans tend to have a narrow horizon for complexity; standards, and especially the standards involved in IT systems, can aid us in dealing with some of the complexity that surrounds us in the world. By providing a platform for coordinated environmental action, IT can help bring about global change in ways that were not previously possible.

In addition to the standards that underlie computing systems, the International Organization for Standardization (ISO) has developed standards for the accounting for and verification of GHG emissions. ISO 14064 provides "specification with guidance at the organization level for quantification and reporting of greenhouse gas emissions and removals" (ISO, 2006). In the process of accounting for GHG and other environmental impacts, institutions of all kinds will require the precision of IT systems to adhere to these environmental standards.

The need for standardization can be seen from two sets of statistics cited in the *Smart 2020* report, which appear to conflict. First, on page 9: "Aside from emissions associated with deforestation, the largest contribution to man-made GHG emissions comes from power generation and fuel used for transportation" (Climate Group, 2008). Then, on page 14: "Of the total emissions from human activity in 2002, 24% was from the power sector, 23% from industry, 17% from agriculture and waste management, 14% from land use, 14% from transport and 8% from buildings" (Climate Group, 2008; citing International Energy Agency, 2008). These two sentences use different categories, and appear to present different orderings of the key contributors to global emissions. While there may be some agreement between the two statements, it is nevertheless difficult to make useful comparisons without agreement on terms and the boundaries between categories. This concern highlights a profound challenge in standardizing ongoing environmental efforts.

Even successful standards still face the challenge of how to respond to new conditions and new information, and adapt effectively. Standards may lead to calcification, in which entrenched processes begin to block progress. Any viable set of standards should have a way to incorporate new knowledge and ways of operating. Just as the U.S. government has Congress, which is tasked by the Constitution to make laws, thereby enabling that government to revise its own operation, so too should any environmental and technological standards have a mechanism for adapting changing circumstances.

Technologies That Help Humans Think Longer Term

As was discussed in the previous chapter, people often have a fairly narrow view of the world. According to a senior executive for a logistics software provider, "People have a tendency to look at a single element of the system rather than looking at the whole system" (Climate Group, 2008, p. 39). Perhaps the most valuable way in which technology can assist people to live more sustainably is to help us take a broader view of the world around us. Technology can do so in a number of different ways.

First, it can allow us to understand the impacts of our actions more clearly. There is a great deal of information that is unavailable to us or not easy to access that could help us see how our actions impact the world around us. Most of this information is simply not provided because it would take effort on someone's part to deliver it, and there is no incentive (financial or otherwise) to do so. In addition, some information on sustainability may be actively hidden by corporations or other organizations that would be disadvantaged by its public knowledge. However, access to this information could encourage more sustainable living. Arguing by analogy, a study found that simply keeping a food diary can significantly increase weight loss (Hollis & Weight Loss Maintenance Trial Research Group, 2008); therefore, it seems likely that simply keeping a diary of environmental impact could reduce one's environmental footprint. The challenge is that, while calorie information for most foods is easy to find, environmental impact information is not. Despite the lack of universally available environmental impact information, there is nonetheless some information of this kind available, and more is likely to be made public each day. By providing us with more contextual information about the impacts of our lifestyles and actions, IT could help us think and act in ways that reflect an awareness of long-term issues.

Second, technology can help us envision new ways of living. Rather than being stuck in a rut of living the same way that we always have, new technology can enable new ways of thinking about our own lives, and new ways that we could choose to live. These ideas can come from other people sharing insights or from our own meditations on how we live. For example, new kinds of technologies can help us focus on "services and systems, not on things" (Thackara, 2005, p. 4); this shift in emphasis could help reduce the impact of our lifestyles. Helping us feel empowered to think in new ways about reducing our footprints and giving us the tools to share insights with each other can provide substantial environmental benefits.

Technology is well positioned to play a critical role in helping people change the way we think about our lives. Many people in the industrialized world seek to keep up with the newest technologies and take their cues about the future from the technologies that are marketed to them. In addition, while technologists were once branded as pocket-protector-wearing, taped-glasses nerds, they have now achieved a certain geek chic. While "computer programmer" may not rank up with "movie star" or "professional athlete" for high-profile allure (except among certain fan communities), there is a certain sense of techno-utopianism in many industrialized nations, in which technology is seen as an essential aspect of a positive future.

Technologies That Help People Live Longer Term

Beyond helping us think about our lives in more sustainable ways, technology can help us live more sustainably without us even knowing it.

Technology can help make more sustainable the infrastructures that support our lives, thus letting aspects of our lives have less impact without any change or awareness on our part. For example, more efficient power generation can reduce the high percentage of energy lost in the generation process without appreciably changing the way people live, except perhaps for a slight price reduction.

Furthermore, new technologies cause people to develop new behavioral patterns, which may in turn lead to more sustainable lifestyles, without sustainability being the primary goal. For instance, telecommuting and online meetings may have been adopted as ways of saving time and money, but they also lead to a reduction in environmental impact. As evidence of this phenomenon, Don Patterson (2007), a professor and colleague at the University of California, Irvine, saved 122.7 tons of CO_2 by hosting the Pervasive 2008 conference program committee meeting virtually rather than flying all its members to the same location.

It is an open question whether it is sufficient simply to enable people to live sustainably, rather than to think sustainably. Here there is a bit of conflict between ideology and pragmatism. One of the assignments that I give my undergraduate Social Analysis of Computerization course involves the following task: "Choose a technology for decreasing resource use per capita, and develop a plan for causing a billion people to use it." One of the challenges that we often discuss is whether it is better to convince a billion people to want to live sustainably or achieve this broad usage by connecting sustainability to some other incentive structure, such as by making a more sustainable product cheaper as well. Given differences in culture, religion, socioeconomic status, and many other factors, convincing a large subset of the world population that sustainability is an important and practical goal is difficult; developing a technology that is cheaper (and incidentally more sustainable) than the existing solution to a significant problem may be an easier path to large-scale adoption. On the other hand, while it may be easier, taking an approach that relies on market-based, capitalist infrastructure may still be problematic (Dourish, 2008).

It is possible that large-scale awareness of environmental sustainability and the themes underlying it is necessary in order for humanity to respond with sufficient vigor to achieve a state of sustainability in the near future. Just as many people in the world live with an awareness of economics (e.g., whether an action will earn us money), law (e.g., whether an action will get us in trouble), or entertainment (e.g., whether something is enjoyable), it may also become relevant for a large number of people to develop a sense of the sustainability of our actions.

Technologies That Last Longer

In addition to new forms of technology that help us think sustainably and act sustainably, it is also relevant to develop technologies that do what they have always done, but simply last longer. Technology is a locus for rapid change in functionality, which often carries with it rapid change in hardware as well. However, there may be innovation paradigms that have less environmental impact than those currently in place. For example, a shift from focusing on products to focusing on services could reduce the need for rapid turnover in hardware. Corporations would no longer have the same incentive to gain profit by selling products, profiting instead by providing ongoing high-quality service to their customers.

Beyond simple functionality, in much of the world there is a culture that embraces newness and disposability rather than tradition and long-term worth. This trend manifests in advertisements (e.g., mobile phone data plan ads where people using an older service sit around waiting for content to download), movies (e.g., Will Smith drawing a distinction between "old and busted" and "new hotness" in *Men in Black II*), and many other media. There is a long tradition in human cultures regarding heirlooms, though—objects that are passed from generation to generation, gaining value in their longevity. Blevis (2007) points out that enabling technologies to take on heirloom status could enhance their sustainability. In addition, enabling technology to last longer could help move us away from the culture of disposability that exists in many places around the globe.

Supplanting Older Technological Paradigms

In many cases, when a computing technology is supplanted, the old version becomes e-waste, rather than being reused effectively. While it may seem that the end user, the one who discards the device, should be to blame for this waste, it is worth considering that perhaps the device was designed in a way that ensured it would become waste as soon as its immediate usage had run its course. Planned obsolescence and the disposable nature of much technology go hand in hand. An additional shortcoming with computing technologies is that it can be easy to forget about older technologies, leaving them plugged in where they continue to draw power (especially if they are left on, which people and corporations often do). In a 2003 talk, Amory Lovins (2003) proposed the idea of "negatechnologies," suggesting that "we tend to discuss only deploying better new technologies. Equally important is getting bad existing technologies out of use." In some cases, no technology at all is better than what currently exists.

A Different Set of Metaphors

One way to help civilizations break out of existing resource-intensive lifestyles may be to think about new kinds of metaphors that can supplant the current ones. For example, rather than providing a GPS device (a tool like a hammer but that tells us where we are), companies could focus on offering a GPS service (a system like a network of friends that know where we are and can tell us). More broadly, the desktop metaphor may itself contribute to unsustainable living by revisiting wasteful paradigms associated with traditional desk work, such as the wasting of paper; alternate metaphors (such as "a computer is an island" [Tomlinson, Baumer, & Yau, 2006]), might help us use computers in ways that are less constrained by the desktop metaphor while simultaneously fostering sustainability. By shifting our thinking about the systems we use, new metaphors can help many different aspects of our lives become more sustainable.

Labor Productivity versus Resource Productivity

The historical and theoretical underpinnings of technology in general, and IT in particular, also help shed some light on how IT contributes to sustainability. For much of the past 150 years, industrialized societies have focused extensively on enabling a given amount of human effort, such as physical labor, to translate into increased output, such as manufacturing. The ratio of output per unit of human effort is known as "labor productivity." An alternate way of thinking about productivity arises from a focus on resources rather than human labor. The notion of "resource productivity"—that is, the amount of output that can be generated with a certain amount of raw materials—features prominently in a future world where resources are increasingly scarce. In many cases, there is a trade-off between labor productivity and resource productivity. For example, one alternative production method will require more human labor but fewer resources, and another will require more resources and less human effort.

One of the core problems in enabling sustainability is the current emphasis on labor productivity at the expense of resource productivity (von Weizsäcker, 2008). The human-centered approach taken in this book could be seen as ideologically linked to labor productivity and therefore potentially at odds with sustainable goals. Labor productivity and HCC both concentrate on making humans the most important element in the productivity equation. However, extending the HCC model to broader scales of time, space, and complexity may go a long way toward rectifying this situation. Although it could entail an increase in the human time and effort needed in production processes, or else an increase in the price of goods, using an extended human-centered approach to shift the emphasis from labor productivity to resource productivity could help us think about our world in more sustainable ways.

Over the last century, labor productivity has been driven in large part by increases in workers' wages. The increase in the cost of labor caused a need to optimize the system for human efficiency. Now, however, with the increasing likelihood of resource scarcity (in terms of not just oil but lumber and other materials as well), the price of resources is likely to rise, thereby forcing industries to increase the efficiency of use of those resources. Von Weizsäcker (2008) contends that governmental regulations that elevate the cost of resources through taxes (perhaps bringing the costs closer to the true cost to the world) could incentivize corporations to think in more resource-efficient ways, thus leading people to more sustainable processes.

There are many examples of ways that various institutions have increased resource productivity without losing money (von Weizsäcker, Lovins, & Lovins, 1998). Citing a variety of cases from wood-chip insulation to waterless urinals, von Weizsäcker and his colleagues point toward a focus on resource productivity that does not entail people needing to sacrifice wealth or lifestyles in order to reduce resource use. The goal is to preserve the same outputs, in terms of functionality to people, with less resource use. This style of efficiency also generates new opportunities for innovative business models, since processes that consume fewer raw materials also usually cost less.

Technology plays an important role in the balance between labor productivity and resource productivity. Technology in all its forms is a facilitator of human activity. Many forms of technology are resource intensive, though, utilizing the deep reserves of natural resources found across the planet. There are many cases of technologies that enhance both labor productivity and resource productivity in various ways. For example, the invention of automobiles has allowed people to travel more efficiently, thereby increasing our productivity. More recently, hybrid cars have allowed people to go further on each gallon of fuel, hence increasing our productivity in terms of resources. The *Smart 2020* report notes that "China's circular economy approach, which recognises the strategic role of resource productivity, is being developed into law, chiefly because environmental pollution is recognised as constraining economic growth" (Climate Group, 2008, p. 6).

Different forms of technology bias human activity in favor of labor productivity or resource productivity to different degrees. For instance, gasoline motors are powerful tools for enhancing human productivity, but they imply a significant focus on the consumption of natural resources. Web sites that encourage bicycling, on the other hand, would favor resource productivity over labor productivity. Still other technologies, such as optimization algorithms, can enhance both human efficiency and reduce the consumption of resources.

Current computing technologies have developed in a culture centered around labor productivity. The term "computer" itself demonstrates the roots of computing in the reduction of human labor; prior to digital technology, "computer" was a job description for a person who performed calculations, either by hand or using a mechanical calculating device. Enabling computing to happen via digital tools rather than through rooms full of people allowed for a vast increase in the efficiency with which people can act.

One of the key principles grounding the argument presented here is that IT is more consistent with sustainable lifestyles than many other forms of technology because it can help enhance systemwide efficiency rather than concentrating on off-loading human effort to nonhuman factors. Mechanical technologies rely on the creation and movement of physical objects, both of which are resource intensive. IT, though, moves bits rather than atoms, and therefore tends to have lower power requirements than many manufacturing and transportation technologies. (For example, mobile phones with gas tanks have yet to hit the market.)

A core challenge in Green issues is that we humans usually do not want to exert more effort than we have to in order to get what we want. Efficiency is an evolutionarily stable strategy; organisms that waste energy survive less well than more efficient counterparts. Humans are no exception; we tend not do things that are too hard unless there are substantial benefits to be gained. This is not to say that we are purely rational actors. Still, most of our actions are inspired to a greater or lesser extent by the desire for some benefit.

IT helps people find out about better ways to perform tasks and share information that supports these efforts. It can be used to encourage collaboration, and the sharing of both resources and ideas. IT can both reduce human effort and increase the value of that effort without necessitating a corresponding increase in resource expenditure.

In order to understand the wide variety of ways that IT can affect people's actions, it is helpful to consider the many different types of benefits that appeal to people. There are physical benefits such as eating a piece of cake or having one's back scratched. There are social benefits such as experiencing feelings of belonging or affection. There are benefits from feelings of control, such as are found in computer games where pushing a button causes pictures and sounds to happen in response. There are also moral benefits that come from feelings of giving back to the world, and the people and other living things in it. Each of these benefits appeals to varying degrees to different people.

In addition, to understand the impediments to people behaving in certain ways, it is useful to consider the many different attributes of an activity that could cause it to be perceived as difficult. There is the physical effort required. There is the loss of time to do it. There is the lack of knowledge of how long it will take or how many resources it will consume; this variability causes uncertainty, which can be inherently dissatisfying and may prevent advance planning. There is also the possibility that the activity might not have the desired results; failure renders useless any effort that was exerted.

The above paragraphs point to two effective ways to get people to do something: either increase the benefits, or lower the costs so that the benefits become greater than the costs. Grocery stores do this by putting items on sale. The Red Cross does this by making people feel important and good when they give blood, offsetting the needle prick and lightheadedness that accompany the act. Social groups do this by exerting peer pressure, raising the social cost of not participating. While humans are not always rational decision makers, looking at the immediate costs and benefits of a given action can be of some help in understanding why people engage in it or not.

The primary benefits of environmental action accrue over a fairly long time horizon. Thus, understanding why people engage in Green activities at present, and how we might be encouraged to do so more frequently, is important to the future of this effort. As will be discussed in greater depth later, Green IT can contribute to this effort by helping make certain activities more likely to occur by either raising their benefits (e.g., online recipes that help people make tastier food with local ingredients) or lowering their costs (e.g., the Web-enabled coordination of communal cooking, so each person only has to cook every few days).

Green IT can foster sustainable lifestyles by facilitating a shift from maximizing labor productivity alone to considering both labor and resource productivity. A large number of people would be reluctant simply to start exerting a great deal more effort for the sake of the global environment. However, by enabling more efficient individual strategies and opportunities for collaboration, IT may be able to help shift the balance toward resource productivity without dramatically increasing the effort required of humans.

For the past century or more, the industrialized world has placed a great emphasis on the efficiency of workers at the expense of raw materials. This focus has caused an unsustainable perspective to dominate in many cultures around the world. IT, though, with an emphasis on efficiency and collaboration, can facilitate a shift toward longer time horizon thinking, enabled by factors such as a greater attention to resource productivity, and hence move people collectively toward sustainability. The discussion here aids in building an understanding of the mechanisms through which IT can connect with sustainable lifestyle choices.

The resource-centered perspective can serve as a proxy for a long time horizon, human-centered perspective. In a resource-centered model, systems are evaluated in accord with how effectively they use the resources involved. In fact, the world currently uses a resource-centered model; however, human effort is the only resource that is deemed important enough to factor significantly into the analysis. Von Weizsäcker's suggestion of increasing the cost for nonhuman resources provides a potentially viable way to cause those resources to factor into this model. This increase can either come from the top down, via taxation, or the bottom up, via people making individual decisions based on the resources used in each option. In terms of the connection to a human-centered approach, since people need resources to live, tailoring people's present behavior to ensure that there are sufficient resources available for people in the future helps guarantee that a long time horizon is viable.

The Need for Green IT

In the environmental community, there has been a slogan in use for several decades: "Think globally, act locally." Understanding the relationship between environmental issues at broad scales and the actions that people can take to help address them is both challenging and important.

Computing technology is particularly well suited for dealing with factors of scale. Mathematics and recursion work equally well with microscopic, human-scale, or universe-sized phenomena; they apply whether events take milliseconds or millennia. Green IT solutions that work well on the small scale of a community may be distributed online and have an impact on other communities around the world. Conversely, if a government, multinational corporation, or other large-scale institution causes a technology to come into existence, there is much infrastructure in place, from mobile phone networks to the Internet, to help disseminate it so that people can use it in our everyday lives. In addition to helping at both large and small scales, Green IT may also be effective in bridging the gaps between the two, and looking for commonalities and opportunities across orders of magnitude of time and space.

Relevant in Space

As was discussed earlier, the global issues facing Earth occur across large spatial areas. Sea level rising, for example, does not just affect one piece of shoreline. A large part of Manhattan will be under water if certain sea level predictions are borne out (Guggenheim, 2006), and the president of Maldives is beginning plans to purchase a new homeland for the people of his country, in case most of their nation (with a maximum altitude of 2.4 meters) becomes submerged (bbc.co.uk, 2008b). This broad spatial impact is relevant across many other facets of global climatic disruption, pollution, species extinction, and other environmental challenges as well.

Since people are skilled at problem solving on spatial scales that relate to the human body (from a few centimeters to a few kilometers), making environmental sustainability relevant to people often requires some interpretation. For example, a common way that people abdicate concern for sea level rising is to say that, since they do not live on the coast, it will not affect them. IT can make the connection between environmental change and people's own lives more apparent, helping people who do not live near the problem to understand the full ramifications of such an event, become concerned about it, and take action. For instance, online mapping software allows for dynamic scaling in space, which can help people see how the locations of things that are important to them relate to the more directly impacted areas. Social networking, too, can make the connection more explicit, by helping people understand the direct impacts of environmental change on their family and friends. Through these and other mechanisms, Green IT can aid in compressing space and enabling local action.

Relevant in Time

Similarly, environmental issues occur across time scales that are largely beyond the ken of individual humans. CO_2 pollution is leading toward global climatic disruption (IPCC, 2007a), but the impact will take decades or longer for the change to manifest fully. Similarly, the rate of species extinction may not be immediately obvious to most people, since the extinction of a given species usually happens gradually over many years. Like a car that needs to start braking several hundred feet before an imminent collision, environmental processes and human practices have a lot of inertia, and need to be corrected significantly before the impact of those practices are felt.

Developing an international response to issues that are so broad in both space and time requires hitherto-unprecedented coordination. Nations and international organizations must work together to coordinate effective action, and do so in a way that is responsive to cultural differences. Arguments that may persuade an atheist may not persuade a Christian or a Muslim, and techniques that could work in the industrialized world may be ineffective in the nonindustrialized one. Green IT can support global efforts by assisting in the gathering of the best science about environmental change and connecting the meaning of that science to impacts and responses across the globe. In addition, high-level global coordination will need to reconcile with local action; Green IT will be relevant at all scales. There will be many challenges in the effort to build global Green IT solutions. As was mentioned earlier, technologies are designed by individuals, and therefore tend to be more effective at solving problems on the scale of individuals. Nevertheless, an effort to support global corporate, governmental, and/or NGO efforts through IT could produce significant environmental benefits.

Other Green IT innovations can make environmental issues relevant in time. Interactive and noninteractive visualizations can show what happened in the distant past. Predictive simulations can allow people to see into the future, years or centuries from now. Time-lapse photography and the online video distribution of it can help change become apparent on the scale of human attention. While it may be difficult to get people around the world to adopt the Haudenosaunee's principle of considering impacts seven generations from now, Green IT may be able to do its part to extend people's time horizons for understanding the impacts of environmental change.

Accessible in Complexity

The complex interdependencies involved in environmental issues are a third major area in which Green IT can help. Achieving global sustainability will require collaboration across many different disciplines and sectors of society. IT can connect experts to each other, enabling them to exchange data and integrate their ideas, and involve laypeople in many stages of the process, helping them commit to the lifestyle transformations that will be necessary.

IT can also enable stronger linkages between efforts at institutional and personal levels. Scaling across levels of social complexity, thereby facilitating accountability from institutions and a sense of efficacy from individuals, will help break down barriers to integrated collective action. It will take active effort by thousands or millions of people, and passive participation by billions of others for sustainability to come to pass. Understanding how Green IT can make environmental issues relevant in space, time, and complexity can aid this effort.

While Green IT is not a panacea, it may be able to help. Like the old slogan "guns don't kill people; people do," technology won't save the world; people will. New forms of IT, different from the growth technologies that helped us create the current predicament, may be able to support efforts to engage with previously untenable issues.

5 Survey of Green IT Systems

Strategic approaches to climate change will involve understanding not only how to do what we currently do more efficiently, but how we can do things differently. The ICT sector will be in a position to enable new ways of learning, travelling, working and living.
—Climate Group, 2008, p. 57

IT is shaping many spheres of human activity, each of which has an impact on the sustainability with which we live. This chapter provides a broad survey of the ways in which IT is currently impacting the goal of sustainability around the world. As Margaret Bruce (2009) offered at the Greening of the Internet Economy conference, there is no silver bullet for the world's environmental issues; we need "silver buckshot." As IT spreads further, and people and organizations become more aware of environmental concerns, the impact of Green IT systems is likely to expand significantly.

There are many potential ways to organize the range of Green IT systems. One could arrange them via the kind of environmental impact that they have (e.g., reduce CO_2 emissions, support ecological restoration, etc.). One could arrange them in relation to their time scale, spatial dimensions, or complexity (e.g., education has a longer time horizon than installing compact fluorescent lightbulbs). One could arrange them via their social scale (from individual action to multinational coordination). In this chapter, Green IT systems are arranged based on the economic sectors on which they will have an impact.

The *Smart 2020* report presents one model for exploring the use of IT to help civilizations be more sustainable. This report summarized its approach as follows:

The challenge of climate change presents an opportunity for ICT to first standardise (S) how energy consumption and emissions information can be traced across different processes beyond the ICT sector's own products and services. It can monitor (M) energy consumption and emissions across the economy in real time, providing the data needed to optimise for energy efficiency. Network tools can be developed that allow accountability (A) for energy consumption and emissions alongside other key business priorities. This information can be used to rethink (R) how we should live, learn, play and work in a low carbon economy, initially by optimising efficiency, but also by providing viable low cost alternatives to high carbon activities. Although isolated efficiency gains do have an impact, ultimately it will be a platform—or a set of technologies and architectures—working coherently together, that will have the greatest impact. It is through this enabling platform that transformation (T) of the economy will occur, when standardisation, monitoring, accounting, optimisation and the business models that drive low carbon alternatives can be developed and diffused at scale across all sectors of the economy. (Climate Group, 2008, p. 11)

Smart 2020, like many efforts, focuses on GHG emissions. This environmental concern, tied closely to global climatic disruption, is of critical importance, but is still just one of several critical environmental issues. Other key domains (discussed in greater depth in chapter 2) include, for instance, other forms of pollution, overpopulation, and species extinction. Many of these issues are coupled; for example, stopping deforestation to reduce CO_2 emissions simultaneously retains habitat. Nevertheless, they are not identical; lots of small areas of forest, for example, can have similar CO_2 effects to one large block of forest, although these two configurations would have different habitat implications. Due to the growing body of research on GHG impacts of various economic sectors, GHG emissions will provide the metric by which the sections of this chapter will be ordered, allowing them to be presented roughly in order of importance for arguably the greatest environmental issue currently facing the world.

The Gartner presentation discussed in the previous chapter (Mingay, 2007) divided the environmental impacts of IT systems into three categories: first-order, second-order, and third-order effects. First-order effects are those that relate to the direct environmental impact of the production and use of the computational devices and systems; these effects were described in the previous chapter. The content of this chapter will look at systems that fall into the second and third categories: specific indirect impacts of Green IT on other sectors, and broader social or cross-industry changes resulting from Green IT. Specifically, the slideshow offered the following second-order effects, resulting from the integration of IT in a range of domains: travel substitution, transportation optimization, e-business/e-government, and environmental control systems. Each of these issues will be covered in discussions of various Green IT efforts. The third-order effects, considered "long-term socioeconomic structural changes" in the report, involved energy intensity, GHG intensity, transportation intensity, and material intensity; changes at this level are likely to result from broad collections of activities both within and outside the Green domain. Nevertheless, these crosscutting themes will shape the landscape on which future Green IT systems will operate.

Especially when looking at the growing impact of nonindustrialized nations, new production processes and ways of life enabled by IT are critical. As China, India, and many other countries move toward higher-impact lifestyles, ways to make all aspects of human life and civilizations more sustainable will become increasingly important. "Meeting the sheer scale of demand for products and necessary supporting services in emerging markets such as China and India and continuing to deliver the services to increase productivity growth in the developed world will effectively outweigh the adoption of the current wave of efficiency benefits per product or service" (Climate Group, 2008, p. 10). Across different nations, different facets of life, and many different scales of time, space, and complexity, there needs to be both infrastructural improvements and ideological change. This chapter presents a survey of the most crucial topics for Green IT, with three specific areas—IT for environmental education, IT for personal change, and IT for community action—being addressed in greater depth in chapters 6–8. By offering these specific examples, arranged by industry, it may be possible to get a firmer grasp on the diverse applications of Green IT.

Agriculture and Food Production

Perhaps the most problematic issue facing the CO_2 footprint of humanity is the deforestation occurring around the world as people seek more land on which to grow food. Deforestation is a double threat; not only does the act of burning the land contribute greatly to CO_2 emissions but the lack of forest eliminates the forest's previous activity of converting CO_2 back to O_2. IT can reduce deforestation in a number of ways. Web access and improved communication can enable farmers to access information about how to farm more efficiently, and to use methods that strip the soil of its nutrients less severely. Lynn Carpenter, an ecology and evolutionary biology professor at the University of California, Irvine who studies restoration ecology, and I have discussed the possibility of using SMS messages to disseminate farming tips based on her work.

According to the Food and Fertilizer Technology Center for the Asian and Pacific Region (2004), "Proper integration of IT in agriculture can help solve the problem of food crisis, and can contribute to the attainment of sustainable agricultural development in the region." This organization offers a variety of examples of how IT can factor into agricultural sustainability, including the "use of IT to reduce agricultural losses, forecast productivity, and enhance production with proper vertical integration of the production centers and the markets" (Food and Fertilizer Technology Center, 2004). Interestingly, these efforts seem directed primarily at agricultural development (i.e., increasing output), with sustainability being somewhat of an afterthought. Nevertheless, this focus perhaps demonstrates that sustainability is often only valued as long as immediate human needs are met as well.

In other areas of agriculture and food production, greater data storage and analysis helps enable genetic engineering of seeds that produce more food in the same space for companies such as Monsanto (Duvall, 2008). (Monsanto is also implicated in environmentally suspect efforts [see, for example, banterminator.org, 2009; Grunwald, 2002].) In research on cattle farming, IT is being used to "monitor, model and manage cattle behaviour," thereby supporting studies of the carrying capacity of grazing land and helping limit the spread of disease (CSIRO, 2008). As mentioned at the beginning of this book, commercial fishers in Kerala, India, are using mobile phones to improve their efficiency; this technology has helped increase their profits by 8 percent and eliminate their previous 5 to 8 percent waste (Jensen, 2007).

Despite the potential for IT to facilitate sustainable agricultural practices, there are hurdles to be overcome for the broad international deployment of such systems. According to a report by the UN Economic and Social Commission for Asia and the Pacific (2008), IT is a major factor in modern agriculture; implementing agricultural IT systems in nonindustrialized countries, however, can be difficult due to limited access to infrastructure such as telephone service and electricity.

Energy

Energy generation and distribution is another critical area in which IT is having significant impacts. Energy is a particularly well-studied area as well, thereby providing numerous examples of optimization techniques. There are some clear opportunities in the energy domain: "Many companies do not know where energy is being consumed, whether in the manufacture or use of their products and services. Many utilities in the developing world are blind to the consumption and loss of energy. Individual functions and departments rarely coordinate to understand how to pool resources or reduce energy efficiently" (Climate Group, 2008, p. 55). IT could enable greater communication and coordination within and between organizations.

One of the most active areas of exploration in the energy arena involves the development of smart power grids. As mentioned in the first chapter, organizations around the world are replacing millions of energy meters on residential and industrial buildings around the state and exploring ways of connecting them in intelligent, integrated networks. This advanced metering infrastructure allows both consumers and energy providers to monitor how energy is being used; reliable monitoring is a key part of reducing energy use effectively. In addition, smarter consumer power meters can be coupled with renewable energy sources such as solar panels to permit customers to sell excess energy back to the power company.

Research in renewable energy, and especially solar and wind power, has suggested that adding renewable energy to a system can create problems with intermittency compared to the predictable process of generating energy with fossil fuels. Power generators are most efficient when operating consistently near peak output. Wind and sun sometimes cut out when the base of power customers need energy; therefore more effective energy management, enabled by IT, is needed. Smart grids, where supply and demand can be regulated on a much finer level of granularity, can help address this kind of problem. In addition, since large-scale batteries are not yet financially viable, more research is needed in smart battery technology to assist with these energy management needs. Another issue that would need to be considered with large-scale wind power usage is the potential disruption of global wind patterns (Keith et al., 2004).

New energy systems are developed and deployed with the assistance of powerful IT systems. Google has been undertaking an effort titled "Renewable Energy Cheaper than Coal" (RE<C) (Google.org, 2009) to address the rapidly growing power demands of Google's server farms. Research at the National Fuel Cell Research Center suggests that a hydrogen infrastructure for powering vehicles could lead to significant improvements in efficiency and environmental sustainability (Stephens-Romero & Samuelsen, 2009). The ZEV-NET project, also developed at the National Fuel Cell Research Center (2009), seeks to develop IT systems to support zero-emissions vehicles and foster the creation of standardized protocols for fleets of such vehicles. Companies such as infotility.com provide software systems to enhance the efficiency of energy grids. IT innovations are integral to the monitoring and control of new energy systems.

Around the world, IT plays an increasingly critical role in the future of the world's power supplies. In research and development as well as installation and operation, support from tailored IT systems will continue to be an important part of the scaffolding for new forms of power.

Manufacturing

Around the world, people need various manufactured goods to live, and also consume many nonessential goods for luxury or other purposes. According to the International Energy Agency (2007), manufacturing industries are responsible for almost a third of global energy consumption and 36 percent of the world's CO_2 emissions. This report lists "chemical, petrochemicals, iron and steel, cement, paper and pulp, and other minerals and metals" (p. 19) as the key sectors, accounting for over two-thirds of the total.

IT systems are helping to reduce the environmental impact of the manufacturing industry. Companies and other organizations use life cycle assessment, a process enabled by different forms of data processing and visualization, to understand the impact of products over the course of their existence and make all stages of the product life cycle more efficient. This process has traditionally involved tracking products "from cradle to grave," considering every stage from the procurement of the first raw materials to the disposal of the waste after the product's use. The manufacturer Aveda (2009), for example, claims to be the "first beauty company manufacturing with 100% certified wind power." According to Andrew Winston (2008), many companies are using technology to enable greater awareness and openness in the process; Sam's Club has the goal to trace its products "from mine to market," Hewlett-Packard recently released emissions data from all its suppliers, and Wal-Mart is seeking to put CO_2 footprint data on its products. Along these lines, the idea of "extended producer responsibility" (OECD, 2009) takes this process even further, holding manufacturers accountable for the postconsumer portion of the product's life cycle as well. All of these efforts rely on effective IT systems.

Adopting new production processes is sometimes difficult, requiring an interruption in business as usual and possibly a temporary loss of profits. "Chinese factory managers find it difficult to stop producing long enough to implement more efficient industrial processes because they risk losing revenue and competitiveness" (Climate Group, 2008, p. 10). This challenge is one of the reasons why longer-term thinking could be beneficial; by enabling people and organizations to assess the long-term benefits of environmentally minded retrofitting and modifications to existing production processes more effectively, Green IT can incentivize these activities.

In addition to producing manufactured goods in more efficient ways, it may also be possible for civilizations to reduce the overall need for manufactured goods. In his introduction to the *Smart 2020* report, Luis Neves, the chair of the Global e-Sustainability Initiative, proposes that IT can help societies "replace goods and services with virtual equivalents" (Climate Group, 2008, p. 6). This notion has been present in the research community for some time; in the 1990s, the movement from "atoms" to "bits" was an important transition across many facets of everyday life (Negroponte, 1996). Enabling reductions in the manufacturing of physical objects (atoms), and replacing them with equivalent or superior computational interactions (bits), could help lower the global environmental footprint.

Transportation

As the mechanism by which people, goods, wastes, and many other objects move around the world, transportation has a large environmental footprint. Various innovations in IT are allowing many different modes of transportation to become more efficient or be eliminated altogether.

In cars, trucks, and other vehicles, onboard computers facilitate efficiency in many ways—for example, by optimizing engine operation, checking tire pressure, and assisting with navigation. Off-board and onboard computers are also used to enable fleets of vehicles to minimize their collective travel distance and time. For instance, UPS recently improved its overall fleet efficiency by planning routes such that its trucks minimized the number of left turns they made, thereby removing the time spent waiting at traffic lights and burning fuel (Rooney, 2007). Car-sharing programs, enabled by online reservation systems and mobile communications technologies, reduce the need for as many cars in many cities around the world. Researchers have also suggested design principles for navigation systems that help people engage with their environment (Leshed, Velden, Rieger, Kot, & Sengers, 2008).

Intelligent transportation infrastructure supports greater efficiency as well. The traditional IT systems that underlie current roadways—traffic lights and loop detectors—can be improved significantly through congestion management systems, incident detection algorithms, and online interfaces to traffic data—for instance, the California Wireless Traffic Report (Calit2, 2004), which offers automated alerts when drivers' primary routes become congested.

The move toward electric and hybrid vehicles, facilitated by computing systems in both design and operation, also supplies an interesting opportunity, suggested by a panelist at the Greening of the Internet Economy conference. It could be possible to use the batteries in these cars as a distributed "load bank" for managing power, thereby assisting with load balancing in a smart electric grid. Millions of car batteries to store or release power at appropriate times could buffer the power supply and ensure continuous power from intermittent generation sources.

In public transit systems, online schedules help people find out when buses and trains will be operating; this increased ease of use can reduce the disparity in speed between driving and taking those forms of transit. A project by one of my students, Andrew Zaldivar, called the Web Browser Environmental Sustainability Toolkit (WebBEST), improves the online mapping of public transit and cars provided by Google Maps; when WebBEST is installed, it inserts a grid of environmental information into the Web page that is displayed on users' computer when they ask Google Maps for directions, including the time, price, and carbon footprint of each mode of transit. A less direct effect of IT encouraging public transit may result from the ubiquity of notebook computers in many parts of the world; when an individual can work on the bus or train but not in the car, a valuable incentive to take public transit is created.

Airplane travel has been made more efficient in a variety of ways. Computer-aided design of aircraft can minimize drag and improve space utilization. Autopilots can enable greater fuel efficiency while flying. Airline fleet planning can optimize craft usage and minimize downtime. The complexities of both the social and technical challenges of air travel lend themselves to IT interventions.

Finally, human-powered transportation is being affected by Green IT, albeit sometimes indirectly. People may be walking more because of car-sharing systems and online mapping software. There are also bicycle-sharing systems in Washington, DC, Barcelona, and Paris (Becker, 2008). Additionally, people are finding social support for pro-bicycle activism, such as the example of the World Naked Bike Ride at the opening of this book. This type of activism will be discussed in greater depth in a later chapter.

Cutting across various different forms of transit, new mobile IT systems help people manage their own transportation choices. Ecorio (ecorio.org, 2009) tracks the user's movements and calculates the carbon footprint of this transportation. Piggyback (Petit & Petit, 2008) provides support for dynamic ride sharing, helping people find others to share the costs of carpooling.

In addition to helping with actual transportation, IT also supports dematerialization—the substitution of virtual interactions rather than physical encounters—thereby reducing the need for transporting physical objects or people. An abundance of research supports virtual interactions, such as work relating to collaboration at a distance (see, for example, Olson & Olson, 2003) and tools to support it (see, for example, Ishii, Kobayashi, & Grudin, 1992). Besides supporting virtual work, dematerialization can enable people to participate in novel forms of social interaction, e-government, and many other new cultural forms.

As mentioned in the first chapter, telepresence shows significant potential for reducing the need for long-distance travel. According to David Hsieh (2009), Cisco's vice president of marketing, emerging technologies, Cisco has deployed 335 telepresence systems across its various offices, hosting 225,000 telepresence meetings; through these efforts the company has seen a reduction in travel by 10 percent, saving US$250 million and 100,000 tons of CO_2. Research in high-bandwidth data transmission (Smarr, Chien, DeFanti, Leigh, & Papadopoulos, 2003) makes possible the future of this and other forms of long-distance interaction.

Architecture and Building

Buildings and other architectural structures are key consumers of power; mitigating this environmental impact is critical to a sustainable future. Simple forms of IT, such as hotel keys that turn off the lights when the guest leaves the room, have been in use for many years, particularly in Europe. More recently, computational technologies have been deployed to improve the carbon footprint of an assortment of built structures— for example, through smart thermostats to computer-controlled airflow systems. The U.S. Green Building Council offers a "Leadership in Energy and Environmental Design (LEED) Green Building Rating System" (USGBC, 2008). Many characteristics required for LEED certification provide opportunities for IT to enhance the design, construction, and operation of built structures.

To date, a large amount of the governmental policy that has been passed in California has dealt with new construction. IT can play a role in improving new buildings, from smart infrastructure to building automation systems to more effective ways of heating and cooling the structures. Open standards that allow for the interoperation of different technologies are an important part of this process so that, as new devices come into the buildings, they can be integrated into the existing infrastructure.

Despite the policy focus on new buildings, according to a panelist on the "ICT and Smart Buildings" panel at the Greening of the Internet Economy conference, new construction accounts for only 2 percent of the real estate in the state of California per year (Meacham, 2009). Therefore, retrofitting older buildings is an essential part of how IT can reduce the environmental footprint of the built landscape. The U.S. government's 2009 "economic stimulus package" includes US$4.5 billion to make government buildings and military facilities more energy efficient (U.S. Congress, 2009).

Another way in which IT can address environmental issues in through its ability to deliver real-time or near-real-time data about energy usage. Until the new generation of intelligent monitors is installed, electricity meters are often read only once per month; this frequency is not rapid enough to allow people to act to change their electricity consumption patterns effectively. By providing greater resolution, IT enables people to make connections between their immediate behavior and longer-term impact.

Sustainable building practices have been used effectively to demonstrate how people can live and work in sustainable settings. For example, Rocky Mountain Institute (2009) is a nonprofit research center created to "foster the efficient and restorative use of resources to make the world secure, just, prosperous, and life-sustaining." Its headquarters is a 4,000 square foot building in the Colorado Rocky Mountains. At 7,100 feet above sea level and with a temperature that can drop to -47°F, the location is far from balmy. However, the building hosts a crop of banana trees as well as avocados, papayas, and passion fruit. The building's heating is 99 percent passive solar, the utilities are superefficient, and the entire building is a model of how enjoyable living space can be compatible with resource efficiency. The building's energy-saving techniques and technologies paid for themselves within ten months. By demonstrating that sustainability does not need to cost more (and in fact, can save a great deal of money), this building seeks to revolutionize the way living spaces are created (von Weizsäcker, Lovins, & Lovins, 1998).

Beyond individual buildings, the design of entire communities can also enable sustainable lifestyles. For example, locating houses a bit closer together can reduce the overall commute time for most or all of the people in that community. Walkable communities can have profoundly positive environmental impacts (Smart Growth America, 2009), while also having physical health and safety benefits (Sightline Institute, 2009). Shared resources such as communal washing machines can also reduce the total impact of the community, reducing the need for each individual or family to own and maintain their own version of the resource.

Looking further ahead, the city of Dubai has proposed constructing an enormous ziggurat that could house a million people (see figure 5.1). By locating people so close together, commutes are greatly reduced, utilities and other resources are much more efficiently supplied, and many other aspects of everyday life are streamlined. In addition, since the ziggurat would be built with an awareness of environmental issues, its design could be optimized to this end. Computational systems would be integral to the design, construction, and operation of such a community. While the ziggurat may sound a bit like science fiction, some of the other efforts undertaken in Dubai, such as an enormous artificial island shaped like a palm tree and visible from space, give some credence to that emirate's ability to execute plans on large scales. (It is possible that the global economic downturn starting in 2008 may dampen the prospects for some of these bold plans.)

In the area of architecture and building, it is also important to note what a panelist at the Greening of the Internet Economy conference suggested: "The greenest building is no building whatsoever" (Calit2 & CPUC, 2009). According to the panel, in a typical office building, a given space is utilized only 17 to 18 percent of the time (given nights, weekends, and occupants being there only part of the time during the workweek). Perhaps the greatest savings in the building domain would be to develop new ways to improve the utilization of existing space, thereby reducing the need for new buildings.

IT Itself

As mentioned previously, IT creates about 2 to 2.5 percent of global carbon emissions (Mingay, 2007; ITU, 2008); these emissions are expected to grow by about 6 percent per

Figure 5.1
A proposed ziggurat in Dubai (timelinks.org, 2009). © Timelinks. Reprinted with permission.

year through 2020 (Climate Group, 2008). Emerging markets will be responsible for a large amount of this growth; while only one in ten Chinese citizens owns a personal computer as of 2008, seven in ten are expected to own one by 2020 (Climate Group, 2008).

IT systems are helping to mitigate the impact of the IT sector itself. The *Smart 2020* report notes that the spread of liquid crystal displays (LCDs) rather than cathode ray tubes (CRTs), and the move from mostly desktop personal computers (84 percent in 2007) to mostly laptops (predicted to rise to 74 percent by 2020), will contribute to reducing this technology's impact (Climate Group, 2008). In the area of mobile devices, smart chargers that turn off once the device is fully charged could reduce the per device effect of the sector. Additionally (as suggested by Climate Group, 2008; Huang & Truong, 2008), mobile phone service upgrades that do not require an upgrade of the physical device could reduce the production of e-waste and the corresponding carbon footprint.

A major user of energy in the IT sector is the abundance of data centers operated by organizations around the world, including corporations (Strand, 2008) and academic institutions (Keller, 2009). According to IBM's vice president for energy and environment, "Energy use in data centers is growing 12 times faster than worldwide energy use" (Lechner, 2009). A key metric in streamlining data centers is power usage effectiveness (PUE). A PUE of 1 is perfect (i.e., all of the energy is being used to drive computation). According to Bill Weihl, Google's Green energy czar, a PUE of 2 is considered good in the industry, but Google maintains a PUE of 1.2 by building its own servers and data centers, which is one of its competitive advantages. In describing the energy wasted in most data centers, Weihl (2009) explained that "servers waste, for no particular good reason, an enormous amount of power."

Nevertheless, a PUE of 1.2 means that 16.7 percent (0.2/1.2) of energy is being used for purposes other than computation—much of it waste. The GreenLight Project (Ramsey, 2008) at the California Institute for Telecommunications and Information Technology is seeking to address the environmental concerns that arise from modern data centers by developing an instrumented data center that can serve as a test bed for better power management processes. Fortuitously, building new generations of supercomputers relies on better power utilization, so the interests of developers of new kinds of computers are well aligned with the search for sustainable power. Bringing the whole industry toward more sustainable practices is the challenge, starting with examples such as Google's and improving on them. On Google's Green efforts, Weihl (2009) added that "the real issue here is one of scale . . . how do we do it in a way that is scalable and will work for everyone else."

According to an article on Kiplinger.com (Wheeland, 2008), the Consumer Electronics Association has produced a sustainability report that provides an overview of how products are being designed to have smaller environmental footprints. "Among the highlights are companies developing more energy-efficient products, using recycled and easily recyclable materials to build their machines, and cutting back on materials used for packaging" (Wheeland, 2008). Additionally, there are numerous opportunities to make the way people use computers more sustainable. According to IBM's vice president of energy and environment, for example, on average a single piece of data is saved thirty times in a given corporation, on various people's computers, servers, and so on (Lechner, 2009). Data deduplication software could reduce this redundancy and increase efficiency. Various corporations are banding together to address a range of environmental issues, through groups such as the Digital Energy Solutions Campaign (behindthegreen.org, 2009) and the Climate Savers Computing initiative (climate-saverscomputing.org, 2008).

Several IT systems are currently being used in research efforts to make consumer-grade software more sustainable as well. One of my doctoral students, Nadine Amsel, is working on a project in this area titled "Sustainable Software Engineering." This project currently has two main parts. First, she is using Amazon's Mechanical Turk system to collect information about software users' attitudes toward sustainability. Second, she is using a device called a "Watts-up" to measure the power draw of various different versions of software. Taken together, these efforts seek to make software users aware of the environmental impact of the systems they use, and help professional software engineers consider environmental sustainability when they are designing systems and writing code. There are currently a number of different qualities that software engineers tend to agree that good software should have—characteristics such as reliability, efficiency, and usability. Nadine's work aims to add sustainability to this list.

Several other students in my group—Joel Ross, Robert Simpson, Nitin Shantharam, and Michael Riccobono—have been looking at ways in which the move toward distributed sensing technologies can be made more sustainable. They have suggested that the need for sensing devices (and the ensuing power consumption and e-waste) could be reduced by the use of "peer-to-peer sensor sharing." In this model, people in cities with dense populations such as Beijing, Seoul, or New York might be able to have phones that contain only a subset of the sensing components that they might need (e.g., a GPS or pollution sensor), with the users relying on sensors carried by other people nearby to provide the sensed data for the other components via Bluetooth or other local wireless protocol (see figure 5.2). This model of shared sensing may seem less effective than the traditional one, in which every device has all the components it needs; there are certain circumstances, however, in which one's coverage might be improved by a network of surrounding devices (for example, if a person were inside a building that blocked the GPS signal). While there are certainly privacy concerns to this sensor-sharing approach, a survey of potential users suggested that a reduction in price could make this strategy viable.

Other Efforts

IT is facilitating the disposal of many different forms of human waste as well. There are many carbon offset programs that provide people and organizations with opportunities to mitigate their impact (and assuage their guilt) by paying for carbon-negative activities and research. For instance, NativeEnergy (nativeenergy.com, 2008), endorsed by the Sierra Club, builds wind turbines, methane digesters on dairy farms, and solar arrays, and supports a variety of other renewable energy projects. The Nature Conser-

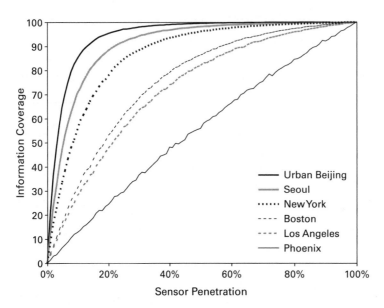

Figure 5.2
Coverage by sensor penetration for select cities in the peer-to-peer sensor-sharing project. In densely populated cities, sharing sensors (e.g., GPS) across mobile devices can allow a relatively small number of sensors to produce a high amount of coverage.

vancy, TerraPass, and many other such entities supply other carbon offset efforts. All of these systems involve elaborate Web sites designed to provide information to potential contributors about the offsets they are purchasing and collect financial support.

E-government also offers an assortment of opportunities for Green IT. For example, Sunne McPeak (2009), president and CEO of the California Emerging Technologies Fund, reports that over the past several years the California Department of Motor Vehicles has moved many services online; as a result, the wait time at its offices has gone from several hours to twenty-one minutes. More important, McPeak underscores that the state therefore did not need to build seven new facilities; every square foot of construction that does not take place is approximately 8.5 tons of GHG that is not emitted.

Online weight loss systems could also provide a significant environmental benefit. Americans use almost a billion gallons of gasoline a year *extra* in order to move the greater body weight of the average American since 1960 (Kloeppel, 2006). Also, given that each calorie of food that is consumed required some multiple of calories of energy to produce—for instance, 40 calories of fossil fuel per calorie of beef protein (40:1 ratio), 39:1 for eggs, 14:1 for swine, 14:1 for dairy, 10:1 for turkey, 4:1 for broilers, or 2.2:1 for plant protein (Pimentel & Pimentel, 2003)—the food needed to nourish this additional weight represents a significant ongoing environmental cost. There are other environmental impacts of obesity arising from the increase in health care as well. Enabling weight loss through online food and exercise tracking could help mitigate these issues.

Ecoinformatics is another area in which IT is having an impact, combining ecologists' research through computational techniques and supporting broad-scale ecological research. One thing that undermines the immediacy of ecologists' impact is the speed with which ecological phenomena operate. It is not uncommon for ecologists to undertake studies on the scale of decades, since that is how long it takes many plant species to grow. Given this long time horizon for ecological research and the immediate relevance of this research to environmental issues, an important way in which IT systems may be able to help is in coordinating and speeding up the process of research and analysis. As an example of the type of project that might be useful in streamlining an individual ecologist's research, consider the ButterflyNet system (Yeh et al., 2006), which provides ecologists with computationally augmented lab notebooks. By enabling researchers to work more efficiently, combining their research findings to provide a broader picture of the state of various ecosystems, and disseminating these findings to wide audiences, IT systems can assist ecologists and other scientists to contribute more effectively.

An illustration of one of the stranger environmentally minded and IT-facilitated projects comes from Australia, where an ecoburial site is being deployed:

Reflecting a worldwide trend towards environmentally friendly burials, the deceased will be buried in biodegradable coffins between gum trees in a protected koala sanctuary. "It's an ideal way of utilizing land and helping wildlife and vegetation," said Kris Whitney, Lismore Council coordinator of cemeteries. "A family can walk around the bushland and pick a site. The body can be oriented in any direction. We promise there will be no internments within five meters. We'll record accurate GPS co-ordinates." Families visiting graves would be lent a satellite navigation device. This will be Australia's fourth "natural burial site" with existing sites in Tasmania, Victoria, and Western Australia. (slashdot.org, 2008)

IT systems are broadening environmental horizons across widely disparate aspects of people's lives, extending even beyond death.

Crosscutting

While the above sector-specific activities and innovations describe some of the most striking ways in which IT is affecting the world, there are many other less apparent ways. Most of these industries operate on an IT infrastructure featuring email, word processing, spreadsheets, simulation, computer-assisted design, and many other computer systems. These crosscutting technologies do not act solely for environmental ends; nevertheless, most or all of the above interventions are enabled to work more effectively by the other supporting IT systems in which they are embedded.

Among other crosscutting ways that IT can improve the sustainability of various industries is to increase accountability. According to Weihl (2009), in most companies "the person buying the IT gear is not the same person paying the electric bill. In many cases, they've never even met." Jim Young (2009), founder and CEO of Realcomm, offered a similar sentiment in a completely different domain, maintaining that there is often a motivational disconnect between the people who build a building, optimizing for low initial costs, and the people who end up living and working in it, who would benefit from greater investment in energy-saving infrastructure. These breakdowns in chains of communication and responsibility, in which different people reap the rewards than the people who make the decisions, provide an opportunity for IT to help people appreciate the true costs of various actions and make choices accordingly.

The Paradox of Green Consumerism

One of the challenges in developing Green products or services is that they nevertheless tend to reinforce the capitalist, market-based system within which they operate. A great deal of effort is exerted across many aspects of the global civilization to ensure the continuation of economic growth. Unfortunately, it is not clear that perpetual economic growth is consistent with environmental sustainability; in fact, many claim that it is not (see, for example, Speth, 2008). By working within capitalist markets, the impact of new systems may ultimately hit a limit beyond which they can no longer move humanity toward a sustainable future.

As an example of this paradox, consider the screenshot of a Web page in figure 5.3. This image shows an article in *New Scientist* titled "Why Politicians Dare Not Limit Economic Growth." The article discusses how advertising is used to enable people to reduce their environmental footprints, but that "the one piece of advice you will not see on a government list is 'buy less stuff.'" Next to this article is a large, color advertisement encouraging users to acquire a new mobile phone every year. Politicians and corporations (such as Reed Business Information Ltd., the company that publishes *New Scientist* magazine) in much of the industrialized world are firmly embedded in a consumption-based society; to oppose consumption on a large scale would be unfeasible in practice.

Many entities engage in a practice that has been termed "greenwashing": making more of a candidate's, product's, or service's green benefits than it deserves, for the sake of selling it more effectively. There are some ideological advantages to the rise of

Figure 5.3

Juxtaposition of article decrying consumerist advertising with a large advertisement suggesting everyone needs a new phone every year. Reprinted with permission.

greenwashing; as people buy more products advertised as Green, they take on identities more firmly rooted in environmental issues and may drive the market to address sustainability more effectively in the future. Dubious assertions about the environmental benefits of various objects or people, however, may cause a backlash, undermining the apparent value of other environmentally oriented efforts and making people skeptical about any future claims. While "Green guilt" (the feeling that one *should* take some action for environmental reasons) can be a motivator in some circumstances, unfounded claims of Greenness can cast a specious light on all environmental undertakings.

Still, while there may be a point in the future where progress- and growth-based market economies make innovations within those systems useless or worse, at present there is a great deal of inefficiency that can be optimized and new lifestyles that can be facilitated to make the "burn rate" of capitalist market economies significantly less than it is now. There is also an unanswered question regarding what kind of economic system would be both viable and enabling of global sustainability. Until that system is found and the revolution comes, there must be efforts to minimize the impact within the civilizations currently in place around the world.

This chapter has sought to provide a summary of some of the ways that IT is helping to reduce humanity's environmental impact. All of the projects described above use IT in some way to address one or more of this planet's environmental issues. There are many other projects operating around the world, some used by millions of people, and some by just a few. Ultimately, all of these projects assist people to think about the future on a longer time scale, understand global issues more clearly, and manage the complexity of relevant issues more effectively. By helping to broaden our horizons, these and other IT systems can make an important contribution to human efforts to restore the viability of the habitats in which we live.

One day when I was in graduate school, I was walking along a paved bicycle path near my Davis Square apartment in Somerville, Massachusetts, on the way to the T station (the Boston area subway). A father and son were walking a few yards in front of me. The boy was about four years old. He was running back and forth across the path, looking under rocks and investigating things. I saw him find something small, pick it up, and carry it over to his father. I heard the father say, "Oh, you found a snail!" I could feel a life lesson about to ensue. "Let's see how far you can chuck that snail, Bobby!"

This snail story provides a lens through which to think about how children develop broad perspectives regarding the natural world and their place in it. In my own life, if I had run up to my parents at the age of four with a snail, they would have had different responses. My mother might have said, "Oh, you found a snail! Let's go put that snail back where it came from, because it probably has a family and they might miss it." On the way, a short biological lesson on snails would probably have transpired. Had I brought the snail to my father, he might have said, "Oh, that's a snail. Why don't you go set it down in the grass, and I'll tell you about a time your grandfather and I found some snails." Then he might have told me about a time he and his father had been fishing, and there were lots of snails around. Other parents might have additional ways of responding to the common experience of being brought a snail by their child, reflecting their religion ("You know, God created that snail"), thoughts on hygiene ("Yuck, go put that down and wipe off your hands!"), or sense of humor (by telling a joke about how the snail's car has a big S painted on it, so when it drives by people say, "Look at that S car go!"). There are different lessons in each of these approaches. Sports are of central importance to some people, biology and environmentalism are crucial to others, and family history to still others.

The perspective that we hold regarding our relationship to the living and nonliving things around us is critical to how we view our role in the future of the world. Humans learn about the world and our relationship to elements in it in many ways. Environmental education is an important part of this learning, from formal educational pro-cesses such as classroom instruction, to informal learning opportunities such as those found in science museums or on nature walks. Fostering an understanding of environ-mental themes is a key component of enabling a large-scale response to the challenges currently facing Earth.

Educational activities are often seen as a way to "broaden people's horizons." In dif-ferent ways, history, science, literature, and many other disciplines expand students' ability to think about scales of time, space, and complexity beyond their own direct ex-perience. This chapter seeks to explore ways in which IT currently impacts the domain of environmental education and the potential ways in which it could have an impact on this field in the future. An examination of the role of IT in environmental education may be helpful in an effort to understand the details of many people's first exposure to environmental issues.

Education and Learning

In order to understand the role that IT can play in education, it is useful to start with a brief discussion of what education is. Education has been described in a variety of ways. The *Oxford English Dictionary* offers the following definition: "The systematic in-struction, schooling or training given to the young in preparation for the work of life; by extension, similar instruction or training obtained in adult age" (OED.com, 1989a). Citing *A Concise Dictionary of Education*, a 1997 report to the U.S. Department of Educa-tion references the following definition: "1. Any process, formal or informal, that helps develop the potentialities of human beings, including their knowledge, capabilities, behavior patterns, and values" (Libutti, 1997). Going back to Latin roots reveals "*e +duco*": "To lead out."

These definitions suggest two key attributes of education: a learner, and some other entity or structure that helps an individual engage in certain kinds of learning. There-fore, education will be defined here as "the process by which one or more individuals or institutions fosters certain kinds of learning in an individual."

Learning, in the broadest sense, is central to many of the projects and themes dis-cussed in this book. This chapter deals specifically with education, which involves both a learner and some structured mechanism for "teaching," rather than with learning in general. Education affects not only how one thinks about the explicit topics involved in the learning process but also one's behavior, worldview, and identity. IT can fa-cilitate both sides of this process, helping teachers consolidate their ideas and present them in a way that supports learning most effectively, and helping learners engage with the educational process in as productive a way as possible.

Formal and Informal Education

Education takes place in a variety of settings and is facilitated by a range of social structures. Schools are perhaps the best-known framework for education; nevertheless, education takes place in other social and community contexts as well. Education research has long recognized the ways in which informal, home, and community-based learning can support and/or conflict with school-based teaching practices. Environmental learning can occur in both formal and informal contexts.

Formal Educational Contexts

There is a range of institutional entities that support formal educational processes. A key characteristic of this type of education is ongoing direct contact with trained teachers in a structured setting.

Various governmental and intergovernmental efforts seek to provide guidance on the processes and materials that are adopted in formal educational settings. For example, UNESCO (2007) is the United Nations' agency with the goal of improving worldwide education. In particular, UNESCO has an initiative that relates directly to sustainability—the Education for Sustainable Development effort. As its site reports, "This educational effort will encourage changes in behaviour that will create a more sustainable future in terms of environmental integrity, economic viability, and a just society for present and future generations" (UNESCO, 2005). Several U.S. governmental agencies have also teamed up to create the GLOBE program, a project seeking to engage students around the world in environmental education through citizen science projects and other efforts (globe.gov, 2009). The GLOBE program provides an array of IT support for this learning; for example, the program's Web site allows students to upload environmental measurements, submit digital images, and access relevant online resources.

Other large-scale organizations also seek to help supply overarching structure to educational undertakings. The U.S. National Research Council provides "National Science Education Standards" (National Academy of Sciences, 1995) that specify areas that all students in the U.S. educational system should know, according to age and grade level. The science that supports an understanding of environmental issues pervades many aspects of these standards. For example, the Life Science Standards include the following topics at each grade level:

• *Kindergarten through fourth grade:* "Organisms and environments."

• *Fifth through eighth grade:* "Populations and ecosystems; Diversity and adaptations of organisms."

• *Ninth through twelfth grade:* "Interdependence of organisms." (National Academy of Sciences, 1995)

In addition to the broad scientific underpinnings, there is also a specific section titled "Science in Personal and Social Perspectives Standards" that includes the following topics:

• *Kindergarten through fourth grade:* "Changes in environments."

• *Fifth through eighth grade:* "Populations, resources, and environments."

• *Ninth through twelfth grade:* "Population growth; Natural resources; Environmental quality; Science and technology in local, national, and global challenges." (National Academy of Sciences, 1995)

California's standards offer:

• *Kindergarten:* "Students know how to observe and describe similarities and differences in the appearance and behavior of plants and animals."

• *First grade:* "Plants and animals inhabit different kinds of environments and have external features that help them thrive in different kinds of places"; "both plants and animals need water, animals need food, and plants need light"; and "animals eat plants or other animals."

• *Fourth grade:* "In any particular environment, some kinds of animals survive well, some less well, and some cannot survive at all."

• *Sixth grade:* "Organisms in ecosystems exchange energy and nutrients among themselves and with the environment."

• *Ninth through twelfth grade:* "Students know diversity is the sum total of different kinds of organisms and is affected by alterations of habitats"; "students know how to analyze changes in an ecosystem resulting from changes in climate, human activity, introduction of nonnative species, or changes in population size"; and "students know how fluctuations in population size in an ecosystem are determined by the relative rates of birth, immigration, emigration, and death." (California State Board of Education, 2004)

Formal education often provides a society's baseline of educational material. The content of formal education standards must be balanced with the standards and curricular demands of other subject areas, and must be approved before they can serve as guidelines for classroom curricula. Therefore, such content may not reflect all perspectives, and may not be able to respond rapidly to current events. The fact that environmental topics are making their way into the canon of school-based content is heartening; however, informal educational settings may be more appropriate for the most cutting-edge content in this area.

Informal Educational Contexts

Education that takes place out of school provides opportunities for learning that may not be appropriate or viable in formal educational settings. Venues for informal education include science centers, after-school programs, community organizations, and homes. The structural entity offering the instruction may vary, from museum exhibits to librarians to parents themselves. Additionally, the learner may be a school-age student or a person at any point in their life ("life-long learning").

As testimony to the importance of informal education, NSF offers a substantial grant-making program called Informal Science Education (ISE) to support these kinds of learning opportunities. According to NSF:

Informal learning happens throughout people's lives in a highly personalized manner based on their particular needs, interests, and past experiences. This type of multi-faceted learning is voluntary, self-directed, and often mediated within a social context (Falk, Dierking, & Foutz, 2007), it provides an experiential base and motivation for further activity and subsequent learning. The ISE program invests in the development of experiences that encourage informal learning in science, technology, engineering, and mathematics (STEM). It promotes public engagement with and understanding of STEM through such means as exhibitions, media projects, emerging learning technologies, and educational programs. ISE-funded projects reach audiences of all ages and backgrounds across the nation in museums, theaters, community centers, in virtual environments and many other settings, including outdoor environments and their homes. (U.S. National Science Foundation, 2006a)

Informal education is well matched to environmental themes for at least two main reasons. First, because private institutions tend to be able to change more rapidly than public ones, informal education can enable a faster response to global changes. As was discussed in chapter 2, the known environmental issues facing the world are becoming worse; in addition, there may be additional environmental concerns that have not yet been discovered. Therefore, the ability to respond rapidly is of significant value.

Second, because many kinds of informal education are not held to the same standards of governmental regulation, it is possible that informal settings could be a better context for learning about potentially controversial topics relating to environmental sustainability. Due to the social forces that must be satisfied in order to change formal educational policies, from science to economics to religion, certain ideas may be prohibited or considered inappropriate for formal settings. Informal settings are thus better for encouraging learning about topics that may be contested by economic stakeholders or other constituencies.

Interactions

Formal and informal educational systems can interact together in a number of ways. For example, grade school children may learn about recycling in the course of their formal education; thereafter they may come home and teach their parents about it—a form of informal education. Alternately, a child's understanding of physics developed while interacting with the informal education presented by science museums could become foundational for learning in a formal physics classroom years later. Sometimes different forms of formal or informal education may clash. For instance, the current debate over evolution versus intelligent design demonstrates that the way current science is taught to a class of high school students does not reconcile well with the way in which Christianity is taught to a congregation.

IT affects the way in which these formal and informal systems interact. For example, students doing their homework can consult informal educational Web sites or use camera phones to share their trips to science museums during class (e.g., show-and-tell). Technological systems find their way into many aspects of society, and both formal and informal educational contexts are no exception.

Educational Challenges

Fostering learning about environmental issues and related scientific fields offers a number of significant challenges to educators. Ecology, for example, is a field of study that pertains directly to environmental issues. The National Science Education Standards discussed above (National Academy of Sciences, 1995), which include education about ecology, recommend the use of discovery-based methods for science education. Discovery-based methods share much in common with the scientific method (e.g., hypothesis formation, experimentation, and analysis), and hence are well suited to most science learning. While many scientific fields offer exciting opportunities for interactive discoveries, however—such as explosions and other reactions in chemistry, or objects in motion in physics—ecology is more challenging for children to interact with in a traditional learning context. The ideas related to ecology are often complex and abstract, involving changes that occur over large amounts of time and space. Different habitats span across many kilometers, and environmental change frequently happens over years and decades. Developing a discovery-based educational program around ecological and environmental ideas is therefore more difficult than other kinds of science education.

To help students learn ecology effectively, educational approaches would benefit from enabling students to explore and interact with natural systems. Most children, though, have limited access to the natural settings that would allow them to study eco-logical ideas effectively. Even if students have access to the outdoors, the ecosystems in their geographic area represent a small subset of all possible habitats. Learning about the relationships among different habitats (e.g., rivers and floodplains) and ecological themes that span many different habitats (e.g., predator/prey interactions) might be improved by direct access to multiple ecosystems. Nevertheless, giving these students access to many different ecosystems is logistically challenging and prohibitively ex-pensive.

Even for students with ready access to natural systems, it may not be viable for them to visit on multiple occasions. Since change occurs so slowly, with regard to the time scales of human activity, returning in a decade to see how the ecosystem has trans-formed is not a viable option for most human learning. The long time spans across which many ecological phenomena operate make it difficult to engage with ecological themes in most educational settings.

Additionally, one of the main mechanisms through which science is taught is via student-led experimentation. However, while it is perfectly fine for students in a phys-ics classroom to measure Earth's gravitational pull, it is potentially problematic for students in an ecology education setting to experiment with living organisms and real-world ecosystems. Such experimentation could raise an array of legal, moral, and ethi-cal problems, from site selection to the choice of intervention to the rehabilitation of failed experiments.

Because of these and other challenges, facilitating ecology education through ex-periential and discovery-based methods can be impractical using traditional tools and techniques. However, as will be described in the next section, IT can help to address many of these issues.

Role of IT

IT is enabling a rapid rise in many new forms of education. Several characteristics of IT make it particularly appropriate for environmental education. While there is no full replacement for direct interactions with real ecosystems to engage learners with envi-ronmental themes, IT-enhanced education has a number of advantages that could be useful in supporting learning around these complex ideas.

First, IT is helpful in its ability to simulate real-world systems. Although a simulation may not have the full complexity of the real-life phenomena on which it is based, the necessary simplification can be beneficial to the learning process. By limiting the com-plexity of the system, a simulation may be able to make the content more accessible to learners and allow them to focus on key aspects of the phenomena. This ability to enable emphasis at specific points may be particularly valuable for educational systems.

Additionally, IT can help to compress and expand both time and space. Years in a simulation can elapse in seconds of real time, for instance, and kilometers can be condensed down to a few pixels. Graphical visualizations can present large bodies of data in the space of a single screen. The ability to manipulate and display information at rates significantly faster than the flicker fusion rate of the human visual system (that is, the rate at which people perceive a series of images as a continuous animation) causes IT systems to be well suited to displaying phenomena across broad continua of time and space.

IT can also scaffold human interaction and understanding through various different media. Allowing learners to act on a system and see the result is an integral aspect of discovery-based science learning, and is consistent with the scientific method. While it is a current topic of debate in educational research communities about how well learning through simulation transfers to the real world, simulation may provide the best option for gaining a robust understanding of complex systems that operate over long time horizons, such as those studied in restoration ecology.

IT is useful for education in part because interactive media appear to have a great appeal for many people, especially children. Interactive systems can engage people with complex content, as can be seen by the growth of the video game industry. Many children spend dozens of hours per week voluntarily playing games, browsing the Internet, and interacting with other forms of IT. Traditional educational systems, on the other hand, often struggle to garner children's enthusiasm.

Looking at broader, ideological questions, IT-based systems are beneficial because interacting with IT does not carry the same kind of legal, ethical, and moral concerns that real-world ecological manipulations might have. Eliminating an endangered species from a real-world habitat is reprehensible; doing so in a simulation is merely educational. IT provides opportunities to run tests and explore alternatives that would be ideologically or physically untenable in real life.

IT-Based Environmental Education Projects

Over the past several decades, a variety of efforts have been undertaken to enhance environmental education through IT. While it is not practical to review all of these projects, this section seeks to provide a number of examples that help to define the breadth of such endeavors.

Classroom-based environmental education is facilitated by IT in a number of ways. Tom Moher's research group at the University of Illinois at Chicago has several projects that illustrate one new form of IT-enabled classroom learning. The HelioRoom project attached several monitors to the walls of a classroom, with circles of various colors moving slowly across and between the monitors to represent different planets, thereby assisting students to understand planetary motion (Thompson & Moher, 2006). The RoomQuake project used monitors to display readings from virtual Richter scale meters, each displaying a different reading when a virtual earthquake occurred, thereby permitting the students to triangulate the quake's location (Moher, Hussain, Halter, & Kilb, 2005). The WallCology project situated a virtual ecosystem on the walls of a classroom, with several monitors serving as portals into that ecosystem through which students could collect data about different species (Moher, Uphoff, Bhatt, Silva, & Malcolm, 2008). Together, these three systems demonstrate a novel way in which technology can be integrated into formal classrooms over multiple class periods, as a means of broadening students' understanding of various complex phenomena.

Another project, IT-E3 (2004), undertaken at the University of California, San Diego and nearby grade schools, seeks to encourage 12- to 15-year-old girls to become engaged with science and engineering through environmental issues via hands-on curricula as well as an online game. In addition, many college and university courses in environmental studies feature IT resources prominently, from simulation to project management to impact evaluation. These in-class uses of IT can strengthen existing formal educational structures.

Outside of school, the Web sites of PBS (PBS.org, 2008), the BBC (bbc.co.uk, 2008a), the EPA (U.S. Environmental Protection Agency, 2008b), and various other organizations (Kaboose.com, 2008) feature a variety of online interactive pages pertaining to nature-based and environmental themes. Computer games such as Sid Meier's Civilization IV (Meier, 2005), Adventure Ecology (adventureecology.com, 2009), and Genius Task Force Biologie (Radon Labs, 2008) provide players with opportunities to explore environmental themes such as sustainability, global warming, and ecological restoration through interactive play experiences. These projects are targeted at a broad audience that spans different age groups and backgrounds, covering topics from pure science to social activism.

Academic researchers and museum exhibit designers have also developed participatory simulations around various topics for out-of-school settings. For example, the Environmental Detectives project led by Eric Klopfer at MIT's Teacher Education Program allows children to play the role of environmental engineers trying to track down a toxic waste spill as they explore an augmented reality by using handheld devices (Klopfer, Squire, & Jenkins, 2002). Participatory simulations allow participants to learn about environmental topics by interacting with a system and other participants. Through this process, they can begin to develop an understanding of the complex problems involved in environmental issues by interacting with a simplified version designed to scaffold their explorations.

Example Project: EcoRaft

EcoRaft (Tomlinson et al., 2006), a participatory simulation created by my research group and our collaborators at the University of California, Irvine, provides a case study of the way in which IT may affect present and future forms of education.[3] The project was a collaboration between my research group of computer scientists and artists, and a group of ecologists led by Professor Lynn Carpenter. It was developed in consultation with educators from Discovery Science Center, a science museum in Santa Ana, California, several miles from the University of California, Irvine.

EcoRaft allows children to learn about the science of restoration ecology by playing the role of ecologists working together to restore a virtual rain forest (see figure 6.1). In this interactive exhibit, three computer-animated "virtual islands" of habitat are arranged around a central space, each displayed on a computer monitor. Each island is inhabited by several different animated species of animals and plants. Children interacting with the exhibit can use three tablet-based personal computers (tablet PCs) as virtual collecting boxes to carry virtual organisms between the islands. The exhibit provides children with the opportunity to see how different species interact and watch the impact of new species on existing habitats. By allowing them to explore these and other ecological phenomena, EcoRaft helps children engage in discovery-based learning acts to figure out for themselves how they might undertake environmental restoration.

During the development of the project, the designers worked with the ecologists to decide on several key principles for the exhibit, so that the simulation would reflect current research in restoration ecology (Carpenter, Nichols, Pratt, & Young, 2004). The final list of principles included:

• Biological communities may be easier to destroy than to rebuild.

• Restoration requires an understanding of the local ecology.

• Conservation of native species and their habitats is important to restoration.

• Restoration is a collaborative effort.

• Humans can have both negative and positive impacts on ecosystems. (Pratt, Carpenter, & Tomlinson, 2006)

Figure 6.1
Children work together to restore virtual ecosystems in the EcoRaft exhibit.

These principles were intentionally modeled after items in the National Science Content Standards for children ages 6 to 14, the target population for the exhibit.

To offer a sense of the vision behind the project, the following paragraphs describe the expected exhibit usage, drawn from a previously published paper (Tomlinson, Baumer, Yau, Carpenter, & Black, 2008):

Sara walks into the EcoRaft space and looks at the three flat screen displays in front of her, all showing images of a lush, green, thriving rain forest with the sounds of birds chirping and a light rain falling in the distance [see figure 6.2]. Each one of the 26-inch LCD monitors is connected to a desktop computer, placed beneath the kiosk on which the monitor rests. As she approaches one screen, she notices that there are small creatures flitting from one plant to another. Sara sees that the creatures in the scene are hummingbirds, as one flies up to the foreground of the screen.

Over her shoulder, Sara sees Thomas run up to another one of the islands. After peering at the screen for a few moments, Thomas sees a button in front of the display labeled "Farm." He decides to press this button, but to his surprise all of the plants and animals disappear, and the rich brown soil on the island turns a sickly reddish color as all the nutrients fade from it. "Oh no!" Thomas cries, "what happened?" Sara walks over and draws his attention to the on-screen text display and accompanying voice recording, explaining that the island has been deforested from over farming. The guidelines on the screen say to use tablet PCs to bring species from the other islands in order to restore this one. Looking around, they see a table with several tablet PCs on

Figure 6.2
In EcoRaft, the virtual ecosystems become more lush as the restoration progresses.

it with a sign showing that they can be used to move species. Sara and Thomas decide to restore the island together.

On the table are three different tablets, each for carrying a particular species. Two are seed boxes, one of which carries seeds for the Erythrina species, also called the coral tree; the other box carries seeds for the Heliconia flower. The third tablet is a birdcage used to carry hummingbirds. Sara decides to pick up the hummingbird cage, while Thomas chooses the Erythrina box. The two walk over to one of the thriving islands, and they notice that this island has no farm button. The island is labeled "National Park," and Sara points out that you probably are not allowed to farm on National Park. Thomas says this is good, otherwise people might farm everywhere and there would be no rain forest left.

Sara lifts the tablet up to the screen, as suggested by the diagram above the table, saying, "Come here, hummingbird, jump." After a moment, a hummingbird flies up to the front of the screen, vanishes from the large LCD, and reappears on the tablet. After watching Sara pick up a hummingbird, Thomas lifts his tablet up to the island. A seed falls off of one of the coral trees and bounces into his box. He notices that, when he tilts the box, the seed rolls around in the box, just as a physical seed would. Seeing that, Sara tilts her tablet too, and the hummingbird responds by flying toward the highest point in the cage.

The two now carry their organisms over to the deforested island. Sara starts by holding the birdcage up to the island, and the hummingbird quickly zips off the tablet and appears on the LCD screen. However, there is nothing growing on that island; the text display and voice over tell them that there are no flowering plants on which hummingbird can feed. After flying around for a time and finding no food, the hummingbird turns and flies away. Thomas suggests that Sara plant a Heliconia flower for her hummingbird. Sara returns the hummingbird cage, picks up the

Heliconia box, gets a seed, and plants it on the deforested island. The seed lands firmly in the soil, but, as the system soon states, there are not sufficient nutrients in the deforested area for the Heliconia to grow. Sara turns to Thomas and suggests that he plant his Erythrina tree, which he does. The tree quickly grows to a large size, and the children can see a ring of brown, nutrient-rich soil emanating from the base of the Erythrina. Within a few moments the system notifies them that the Erythrina has added sufficient nutrients to the soil, and the children watch excitedly as the Heliconia seed germinates and grows into a flower. Sara and Thomas rush over to pick up the birdcage, collect a hummingbird, and bring it to the newly restored island. The two release the hummingbird and laugh as it begins contentedly feeding on the Heliconia flowers they planted. The two children try moving the species around in different combinations, and teach several other children who come over to the installation how it all works.

Implementation Details

The infrastructure of EcoRaft is an example of a growing area of IT known as *multi-device systems*—systems that involve several collocated computing devices working together to provide a coherent experience across all of them. The multi-device system on which the EcoRaft exhibit operates is based on several main implementation elements—networking, graphics, sound, embodied mobile agents, real-world sensing technologies, and a novel interaction metaphor.

Networking   EcoRaft uses infrared (IrDA) and wireless Ethernet (Wi-Fi) signals to communicate among the component devices. IrDA is the same technology utilized in television remote controls. Wi-Fi is the technology typically used to enable wireless Internet access in office buildings, cybercafés, and homes.

There are several characteristics that make IrDA well suited to the EcoRaft system. First, it is relatively short-range, thereby enabling the system to sense the proximity of the devices to each other. Second, it is directional, requiring the two devices to be facing each other with a line of sight between them. This directionality allows the system to have an awareness of the relative orientation of its component devices and thereby coordinate the graphics between those devices, creating the appearance of graphical continuity between them (see the "Graphics" section below). Third, IrDA came built-in with the tablet PCs, thus preventing the need for external attachments that might break easily. IrDA adapters were also readily available for the monitors.

One drawback of IrDA is its speed. While it is fast enough to transfer small amounts of information, it is not fast enough to permit significant graphic information or other data to be passed across it without a human-perceivable lag. Wi-Fi is much faster, but lacks the proximity and directionality characteristics of IrDA. As such, in EcoRaft the devices identify each other by exchanging IP addresses via IrDA and then switch to

Wi-Fi to conduct the bulk of the necessary interdevice data transfer. This hybrid approach allows the system to utilize the most effective characteristics of both IrDA and Wi-Fi communication media.

Graphics Once two devices have established communication through IrDA and Wi-Fi, it becomes possible for them to create the appearance of a continuous graphic world between them. This illusion is created by playing the first half of an animation on one screen and the second half on the other, and carefully timing them so that the second half starts shortly after the first. For example, in order for a hummingbird to fly from one of the monitors to one of the tablets, it would first need to animate from its initial position to the foreground of the monitor, and then disappear from the monitor and reappear at one edge of the tablet.

Proximity and orientation are important to the believability of cross-device animation. If the two devices are far apart, a person will not be able to see both of them at once. If the devices are not oriented correctly, the motion on one screen will not appear to be continuous with that on the other screen.

Certain assumptions are made in the EcoRaft exhibit to help the graphics coordinate smoothly across different devices. For instance, the monitors are assumed to be upright, while the tablets are assumed to be horizontal, facing up. These assumptions allow the relative orientation sensed by the IrDA to specify a unique relationship between the two, and ensure that the animation appears to move believably from one screen to the other.

Sound Just as graphics can be coordinated across devices, so too can sounds be played on various devices to enhance the believability of the multi-device virtual world. Specific sound effects, such as a *zip* for a flying hummingbird, transition from one device to another, thereby helping create the illusion that various organisms are moving among the devices.

In addition to specific event- or agent-driven sounds, ambient sounds also arise from each device, thereby creating a dynamic surround-sound system that helps create an engaging experience for participants. For example, one island in the exhibit features audio based primarily on the sound of a waterfall, a second has several distinctive birdcalls, and a third has wind blowing through trees. This multi-device soundscape helps participants orient themselves among the different devices. The sound track from each island also changes over time based on the success of restoration on that island, thereby allowing participants to sense which islands are most in need of restoration.

Synchronization of audio across several devices is an important challenge in the development of multi-device systems. Enabling aural continuity across devices supports the graphical seamlessness of the cross-device animation. In addition, audio provides subtle but valuable cues to the participants about where to focus their attention.

Embodied Mobile Agents   The animated characters and creatures that inhabit multi-device systems are called *embodied mobile agents* (EMAs) (Tomlinson, Yau, & Baumer, 2006). EMAs are a kind of autonomous agent—that is, a computational entity capable of acting in its world (whether a real world or a virtual one) to satisfy its needs and desires. The research field that studies autonomous agents has much in common with artificial intelligence, but focuses more closely on agents as entities that are situated in some environment and the interactions among groups of agents (i.e., multiagent systems).

The hummingbird EMAs serve as the primary "point of contact" for people interacting with EcoRaft. In addition to the hummingbirds, the seeds of several plant species are able to bounce out of the plants and into the collecting boxes carried by participants. EMAs provide an agent-oriented way of connecting the different devices together, enabling people to perceive the devices as parts of a unified system. Seeing an EMA move from one device to another helps people understand that each device is part of the whole, and that they can therefore interact with all of the devices together.

Real-World Sensing Technologies   EcoRaft also features two real-world sensing technologies that enhance the feelings of engagement and believability fostered by the exhibit. The first, accelerometers that are built into the tablets, allow hands-on interaction with the EMAs. For example, when a participant is carrying a tablet with a hummingbird EMA in it, the hummingbird will always fly to the highest point on the screen. If a participant is holding the tablet in both hands, facing up, and raises the right side, the hummingbird will fly to the right; if the participant tips the front down, the hummingbird will fly to the back. In the same way, the seeds of the various plants can be made to roll around in the tablet, exhibiting simple physical behaviors such as rebounding off the walls of the virtual collecting box. This direct interaction with entities in the multi-device virtual world helps to break down the barriers between that world and the real one, potentially leading to an increased transfer of learning.

A second sensing technology is a Webcam mounted above each of the three monitors. When a participant approaches one of the monitors the Webcam above it detects the movement, causing one or more of the hummingbirds to fly up to the front of the screen. In addition, movement by a participant causes a flower hanging in the fore-ground to sway, as if caught in the breeze caused by the participant moving.

Taken together, these two real-world sensing technologies make a stronger con-nection between the world in which the participants live and the world of the hum-mingbirds, thus potentially helping participants to make connections between lessons learned in the simulation and phenomena in the real world.

Interaction Metaphor   The project is based on a novel metaphor that underlies the way people interact with computers. Existing metaphors that direct people's interac-tions with devices include: "The computer is a typewriter," "The computer is a desk-top," and "The computer is a notebook." Interactional metaphors are useful because they allow people to draw on their experience with familiar objects and interactions to inform their interactions with unfamiliar technological innovations (Negropon-te, 1991). In the EcoRaft system, the core metaphor is: "The computer is an island." Along with this metaphor comes the idea that mobile devices are objects that might carry things between islands, such as rafts, boats, or scientists' collecting boxes. This metaphor helped to inform the implementation of the project and frames children's interactions with the technological system in a way that is more grounded in environ-mental themes than previous metaphors. Whereas typewriters, desktops, and note-books are all products of human ingenuity, and hence help to perpetuate a separation between people and other entities in the world, islands are not human innovations, and therefore may help to break down the separation between people and other parts of nature.

Insights about Environmental Education and Technology
EcoRaft has been shown to several thousand people to date, including several hundred children at Discovery Science Center and our research lab. In order to evaluate the exhibit's impact on audiences, the research team interviewed thirty-eight participants, including fourteen children. These interviews, in addition to passive observation of many of the several thousand other participants interacting in small groups, provided some insight into the ways in which people perceived the system. The following para-graphs describe some findings from these studies. (For a full analysis of these inter-views, please see Tomlinson, Baumer, Yau, Carpenter, & Black, 2008.)

One theme that emerged from the interviews involved the idea of self-directed learning. When asked how he learned how to interact with the system, one child said, "Well, I kind of looked at someone else, and when I just came there, I moved the blue button and I saw what happened to the screen, and I knew it was a bad idea to touch it. I am not really sure how I figured the rest." This response about figuring it out by looking at other people was echoed by many other interactions witnessed by the research team, suggesting that EcoRaft provides participants with sufficient scaffolding to figure out how to interact with the exhibit. A number of design elements supported participants' self-directed learning, including periodic graphical and audible feedback (e.g., "The heliconia you planted can not grow in completely degraded soil"), and the positioning of text on the screens of the tablets so that participants "naturally" tended to hold those devices in the proper configuration for the devices to communicate. Enabling children to figure out for themselves how to interact with a system is helpful in ensuring they feel both a sense of discovery and empowered to explore aspects of the system.

A second major theme that arose in the interviews involved collaboration. Each of the children interviewed mentioned collaboration in some way. Collaborating with others was described by one participant as "more fun" and that they "need to work together." Collaboration was one of the key themes that emerged from discussions with the ecologists, and as such, had been built into the exhibit's structure; since each tablet could carry only one species, at least three participants had to work together in order to restore an island fully. Having collaboration emerge in the interviews suggests that the technological and design elements succeeded in encouraging people to work together while restoring the virtual forests.

A final theme that recurred during the interviews was a push toward complexity. When asked how the exhibit could be improved, one child requested "lions and tigers! And jaguars! And snakes! Big snakes! Poisonous snakes! Alligators and crocodiles." This response certainly showed enthusiasm for the idea of ongoing interactions with virtual species, but also pointed to one of the challenges faced by designers of interactive technologies. People often feel a system would be improved by additional capabilities—a phenomenon sometimes called "feature creep" in the design of software systems. Too many additional features make it harder to understand the system as a whole, however, especially for novices. Hence, although there was excitement about adding other species, the designers decided that a smaller number of species would help participants engage with the core ecological themes more effectively in the short period of time available at a science museum than a more abundant virtual ecosystem would. The development team is nevertheless contemplating future versions of the exhibit that might involve some of the charismatic megafauna mentioned above.

IT can facilitate new ways of responding to the challenges in environmental education described earlier in this chapter. EcoRaft provides one example of how new forms of interactive technology can enhance education around environmental themes, enabling participants to experiment with ecology in ways that would be impractical, immoral, or both if they were attempted in the real world.

In terms of practicality, the simulation helps to compress time and space, allowing visitors to interact with phenomena that occur over dozens of years and hundreds of kilometers. The exhibit helps children engage with deep ecological concepts, despite their relatively short attention spans and limited ability to travel in the world, condensing several decades of ecology research in Costa Rica into a five-minute experience at their local science museum. In addition to making it more interesting (cutting out all the "boring" parts waiting for trees to grow), the system also narrows the focus of the experience, guiding learners to attend to the most salient elements of the systemic processes that determine how species interact. By directing the learners' attention, the exhibit helps them learn key lessons in ecology without the often-overwhelming potential for distraction found in most real ecosystems.

In terms of morality, enabling children to experiment with simulations can allow them to learn about the impacts of various decisions without any harmful effects on real-world organisms or ecosystems. While children also need to learn that there is no "reset button" on real ecosystems, having a wider margin of acceptable error is worthwhile when trying to figure out complex phenomena, especially in the earliest stages of the process.

Thus, while there is no substitute for the engagement that could occur while crawling around the rain forest, looking at insects and smelling the dirt, EcoRaft does appear to provide a number of advantages in facilitating students' learning about ecological themes.

EHCC Perspective

The EHCC described in chapter 1 provides a lens through which to understand the environmental impact of Green IT projects such as EcoRaft.

Looking specifically at EcoRaft, the project broadens the horizons of its users, engages people physically with environmental topics, and positions humans at the center of both the world's current environmental issues and the future means of addressing them. The interactive experience extends users' horizons in several main ways. First, the simulation speeds up time, letting participants take action and see results that would take many years to unfold in real life. For example, the growth of a tree takes less than a second in EcoRaft, rather than decades as in the natural world. This condensing of time allows people to learn about the underlying principles of phenomena that occur over decades or longer in reality. By helping people's powerful ability to learn about short-term phenomena focus instead on much longer-term events, EcoRaft extends users' temporal perspective. Similarly, the simulation condenses space by enabling participants to engage with an entire ecosystem in the space of a single room and reduces complexity by abstracting away many of the details of ecological interactions.

Second, the exhibit concentrates on ecological restoration, a process that is ideologically separated from the ecological destruction that is more characteristic of the modern era. If destruction is like a car going forward, and conservation is the process of bringing that car to a halt, then restoration is putting the car in reverse. Effectively, restoration lies on the other side of conservation from habitat destruction. By connecting people to ideas that may only be possible on a large scale many years from now, but that are nevertheless a major need in environmental action, EcoRaft aids in extending the time scales on which people think.

The way the project engages participants' bodies in the interaction is also important to its potential for environmental impact. By centrally incorporating the tangibility and mobility of the tablets in the interaction, the interface helps people learn a visceral connection to the ideas in the exhibit. Also, by fostering collaboration among participants, EcoRaft builds a sense of connection between people by associating a social reward with environmental action.

Finally, by positioning the human participants at the center of the exhibit (both ideologically, because of the key role people play in the virtual restoration, and physically, due to the structural configuration of the three monitors facing in on a central interaction space), EcoRaft helps people implicitly understand the critical role that people play in present and future environmental transformations. Humans created a great deal of the current environmental change, and humans need to take action to make it better. Through the interaction it enables, EcoRaft places participants at the heart of a virtual ecosystem, a restoration effort, responsibility, and the possibility for positive change. EcoRaft is just one example of "the potential of videogames and immersive participatory simulations for engaging children in rich socio-technical contexts, where they address meaningful problems, and through which they can learn and collaboratively experience all stages of scientific inquiry" (Barab & Dede, 2007).

Looking beyond EcoRaft to other educational Green IT projects, there are many different ways in which the EHCC approach broadens people's horizons. IT-enabled formal and informal educational efforts allow students to develop the intellectual grounding to understand the complex interconnections across evolutionary time scales, within and between ecosystems, and among many different areas of study. IT can also facilitate the engagement that will allow students to motivate themselves to continue learning beyond required educational participation. Education of children and adults around the world is a critical part of enabling human civilizations to become sustainable; extending people's horizons can assist us to make informed decisions in all facets of our lives.

In the coming decades, education may develop in a wide variety of directions. Currently in the United States there is an emphasis on education that prepares students for standardized tests. While ensuring that all children in a society share certain capabilities is an excellent idea, the processes currently being used to enact this idea do not always take into account the variety of learning styles, interests, and backgrounds of the diverse children across the country. A more agile and responsive system, run by teachers and facilitated by technology, that permits each student to learn in ways most effective for them could be an improvement over the monolithic educational system currently in place in the United States.

Educational systems of one kind or another are ubiquitous across human cultures. In order for environmental ideas to have longevity, there must be an educational plan through which future generations are encouraged to understand their role in the global ecosystem. IT can have a major role in facilitating forms of education that help people to understand and embrace sustainable ideals.

Education plays a key part in the political process of environmental action as well. As Bruce (2009) commented at the Greening of the Internet Economy conference, "If all good policymaking comes from the people, the people need to understand the science. . . . ICT can facilitate that." Enabling people at all levels—from grade school through university and into their lives as citizens, professionals, policymakers, and educators themselves—to broaden our horizons of time, space, and complexity can help the global civilization respond effectively to global environmental change.

Returning to the snail story that opened this chapter, the informal and formal lessons that children are taught throughout their lives about their relationship with other elements in the natural world have a great impact on their environmental decision-making processes. This chapter has offered ideas on how technology may be situated in environmentally grounded education to support core lessons about our engagement with the world around us.

7 Green IT and Personal Change

My grandfather worked at Scott Paper Company during World War II. In order to save energy and thereby help the United States fight the Nazis, the company had a rule that, if employees were going down two floors or up one, they should take the stairs rather than the elevator. I have sought to use this strategy in my own activities, and sometimes mention it to classes of my students. It makes sense to me (given how much energy it must take to move an elevator up and down a building), and is a kind of personal change that is easy to understand and adopt into one's life. Recognizing how personal decisions and actions connect to major impacts is a kind of broadening of horizons. These connections may be obvious, in which case the effort is primarily one of reminding people of them and overcoming behavioral inertia, or nonobvious, in which case the effort also involves making the case for the value proposition and specifying the benefit of the action. They may involve integration across various facets of one's own life, or integration of one's own life with aspects of other people's lives or with existing or new infrastructure. Encouraging people to enact personal change can be beneficial for several reasons. First, behavioral choices can have a direct effect, reducing the person's environmental impact. Second, taking environmentally sound actions can reinforce people's identities as environmentally minded individuals. Finally, environmental decisions can provide support for other people who might also wish to change their behavior to serve environmental ends. All of these actions help to broaden people's horizons, extending our understanding and impact across time, space, and complexity.

This chapter seeks to understand how IT can help enable personal change. By giving us the tools to understand the realities of our own lives, conceive of how they might be different, and work toward lifestyles more consistent with our ideals, IT can provide a powerful force for personal change.

Audiences

The amount of environmental impact that each individual causes through our every-day activities may seem inconsequential given the magnitude of the problems facing the world. However, when multiplied by 6.7 billion people, the cumulative impact is great. Many people are beginning to live their lives in ways that reflect sustainable ideals; however, many others are unaware of the issues or uninterested in changing their behavior.

The primary group that this chapter will discuss is residents of industrialized nations and in particular the United States, where resource consumption per capita is highest in the world. Still, it is important to note that, with efforts being undertaken to raise standards of living (which are accompanied by increased consumption patterns) around the globe, it is becoming more and more relevant to address audiences in non-industrialized nations as well.

Enacting lifestyle change can be challenging. In most behavioral change, there is significant inertia to overcome at both the personal and cultural level. At the personal level, we each have habits that help us manage the complexity of our lives. Rather than deciding what brand of toothpaste to use each time we go to the store, we buy the same brand we bought last time. Rather than seeking a more efficient way of getting to work, we travel the path that is familiar because we traveled it yesterday. Rather than adopting a new hairstyle, we brush our hair the same way we usually do. There are too many decisions to be made every day for a person to reconsider more than a small fraction of them on any given day.

Culturally, there are many expectations placed on a person by the individuals and institutions with which we interact. Styles of dress, ways of speaking, and many other lifestyle choices are heavily prescribed by cultural norms. While different cultures have vastly different ideas of what constitutes appropriate behavior, within a given culture or community, the expectations of others exert a strong influence on people's everyday behavior.

In order for people to enact lifestyle change, it would be helpful not only for us to have information about the impact of what we are doing and how we might do it differently but also for there to be support for implementing positive lifestyle/environmental change at both the personal and cultural levels. Behavioral change needs to fit into the personal and cultural narratives that surround all people.

There is growing interest in the United States regarding environmental themes. For example, in 2002, "more than 2 in 5 Americans (44%) say they would try harder to be green if more environmentally friendly products and services were readily available to them, and an additional 20% would do so if they thought their efforts would make a difference" (Gardyn, 2002). The first half of this quote points to an important aspect of personal change: in many of its forms, it benefits significantly from improved infrastructure that provides more accessible pathways to sustainable living.

There are two significant hurdles to be overcome in enabling a person to adopt some kind of lifestyle change: convincing them to try it, and getting them to adopt it as a new habit. An old friend had the mantra that doing something once was an experiment, doing it every day for a week was a habit, and doing it every day for a month was a lifestyle. When attempting to take some new action on a regular basis, one is confronted with many different aspects of the change—how it makes one feel over time, how different social partners respond to it, what resources it uses, what wastes it produces, and what new communities it introduces into one's life. These impacts can be different after a month than they are after a day. For example, longer-term phenomena such as feeling healthier or having a new group of friends may not be obvious after one day, instead emerging over weeks or months; nevertheless, they may be powerful motivators for continuing a behavioral pattern.

Given the vastly different cultures and individuals around the world, and the broad range of potential pathways for environmental change, this chapter will be able to address only a small number of them. Nevertheless, it will seek to present a variety of examples, and go into greater depth with one specific project.

Impact

Enabling people to enact behavioral change can have a number of impacts, including direct environmental effects, changes to the person's quality of life, shifts in their likelihood to enact further behavioral change, and influences on their social partners.

The most apparent impact of behavioral change for environmental purposes is the direct effect that the changed behavior has on one or more environmental issues. It may reduce a person's overall carbon footprint, help restore an ecosystem, or limit human reproduction, thereby helping control world population. All sorts of personal behavioral changes can have direct environmental impacts, such as new purchasing habits, driving less, losing weight, habitat restoration, and birth control.

Another impact that does not directly relate to environmental concerns is the effect that such choices have on the person's quality of life. Many environmentally oriented changes help a person feel better physically or psychologically, live more economically, or find new social opportunities. These effects may make it more likely for people to adopt a behavioral change over a longer term. In addition, aspects of improved quality of life such as increased female education lead to lower birthrates (Akmam, 2002); raising quality of life could be an effective way to reduce the growth rate of global populations.

Looking at the international potential of Green IT for personal change, while industrialized countries use more resources per capita and therefore might be the most fruitful targets (from the point of view of immediate environmental impact) for future innovations in this area, it may be that residents of nonindustrialized countries would stand to benefit most from these innovations. Because people in nonindustrialized nations often make less money, and thus have smaller or nonexistent margins of safety in terms of resource surpluses, tools for personal change that can both reduce environmental impact and improve quality of life could be even more valuable to those individuals than they would be to people in the industrialized world. Tools that enhance sustainability and tools that enhance quality of life can go hand in hand.

Adopting one kind of environmental change may also cause an individual to associate more strongly with environmental principles and communities in general. Individuals can gradually be introduced to new identities by means of smaller actions that then pave the way for more substantive behavioral change. This identity shift will then affect the likelihood of further individual change.

Finally, environmental behavioral change can affect the social partners of the person making the change, and that person's culture more broadly. Social support is critical to people changing their behavior; therefore, the act of one person changing their behavior, if done in a public manner, may provide an incentive for others to do so as well. On a large scale, this process enables behavioral change to move through a population, thereby facilitating a cultural shift over time.

Green IT for Personal Change

There are many specific ways in which people find out about, try, and adopt new forms of behavior that have positive environmental ramifications. Most of these efforts do not entail computer-based IT, and instead rely on more conventional mechanisms such as posted signs (e.g., placards above the paper towel dispenser in many public bathrooms encouraging reduced consumption), notes written on paper, word of mouth, and the like. There is a growing collection of Green IT systems, however, that aid people in enacting personal change, sometimes by replacing existing techniques, and at other times by developing new ways to change one's life. This section provides examples of existing IT-based mechanisms for enacting behavioral change.

Information

Many computer-based IT systems exist to supply us with additional information about our lives and the world around us. Much of this information can have environmental ramifications, if people choose to act on it. Information on weather, transportation, consumer prices, and many other facets of life can help people make more environmentally sound choices. Computational systems increase the ease with which we can access large amounts of information, and therefore enable us to use that information to change our lives.

As a brief survey of the vastly different kinds of IT systems that can provide people with an array of potentially useful environmental information, consider the following efforts. Kill-a-watt is a hardware device that goes between any power cord and the wall socket, and reports back on the energy usage of the device that is plugged into it (P3 International Corporation, 2008). By providing this information, the system enables people to find out more detailed information about the power draw of their devices. Happy Meat is an online mapping system that assists people in finding sustainable meat products, thereby enabling them to indulge their carnivorous habits with as little impact as possible (saisriskandarajah.com/happymeat, 2009). TreeHugger is a Web site that collects information about how to live in environmentally sustainable ways (tree-hugger.com, 2009). There are many other information sources, both online and offline, that explicitly or implicitly provide information that enable sustainable lifestyles. Taken together, these information sources can help us make environmentally sensitive decisions across many facets of our lives.

Analysis

Despite the large amount of information available via the Internet and other computational sources, information alone is not always sufficient to help people make behavioral changes. People may not seek out the information, may not trust the information, and may not be able to reconcile conflicting pieces of information. For example, there is a lot of information about the benefits of energy-efficient compact fluorescent lightbulbs; there is a conundrum, though, about when to replace incandescent lightbulbs with compact fluorescents—wait until they burn out, or replace them right away and dispose of the still-working incandescent ones.

The human memory is perhaps the most effective tool for understanding how various pieces of information fit together; nevertheless, human intelligence can be augmented and enhanced by technology. Many IT tools exist to help people interpret and analyze the information that surrounds us. For instance, ManyEyes is an online system that allows people to upload data sets (e.g., global environmental statistics or bodies of text), create visualizations of those data sets using a variety of techniques (see figure 7.1), and share both the visualizations and the underlying data sets with other people (Danis, Viegas, Wattenberg, & Kriss, 2008). StepGreen is a Facebook plug-in created by Jennifer Mankoff and her collaborators to help members track the financial and environmental savings that result from various Green actions (Mankoff, Matthews, Fussell, & Johnson, 2007). Wikipedia, the collaboratively generated online encyclopedia, collects survey articles on a wide variety of topics that provide users with sufficient context to assist with their own interpretation of environmental issues (wikipedia.org, 2009). Microsoft has begun a grant program called Memex, to support the collection and analysis of "Digital Memories." Enabling people to analyze bodies of information to understand how they may be integrated to support sustainable living is critical to the effort to live sustainably on a large scale.

Action

While information and analysis are valuable for developing an understanding of environmental phenomena, in order for humans to live more sustainably we need to take action (and also begin to refrain from actions that are unsustainable). There are many different examples of action-related IT systems. For instance, by using onboard computers that come built into many hybrid automobiles and other vehicles, people seek to drive as efficiently as possible—an activity called "hypermiling" (Gaffney, 2007). Dynamically controlling home heating and air-conditioning by means of smart thermostats can allow for greatly reduced home energy usage. Carbon offset programs also provide a pathway by which to take action, in cases where people cannot change their behavior or do not wish to. In these situations, carbon offsets can enable them to compensate for the damage their behavior is causing.

Figure 7.1
A "Wordle" visualization showing the top fifty words (excluding common English words) used in the "Summary for Policymakers" of the *Climate Change 2007* report (IPCC, 2007b). Created using Many Eyes (Danis, Viegas, Wattenberg, & Kriss, 2008).

A substantial community of people on the Internet have begun to engage in the process of "life hacking" (Thompson, 2005), in which individuals seek to live their lives in accord with their principles and desires as effectively as possible, rather than succumbing to the inertia of how society encourages people to act or how they have acted previously. IT systems exist to support life hackers, from software to track various facets of their lives, to social networking systems to enable life hackers to exchange ideas.

Joseph Stalin is quoted as having said, "A single death is a tragedy; a million deaths is a statistic." This idea—that people can comprehend horrible things that happen to individuals, but somehow are not able to scale this understanding to large numbers—is relevant to the global environmental situation. Seeing a baby seal get clubbed, a frog with three eyes, or a tree being felled can have a visceral impact on people. Seeing a chart that plots the decreasing number of species in the world over the past several decades, however, does not capture the fact that each species that is lost probably has many individual stories to go along with it. We need better methods of presenting data, whether about our own lives or the entire world, in ways that compel people to action.

Mobile IT

While mobile systems do not precisely fit with the previous three categories (information, analysis, and action), they are particularly relevant with regard to new kinds of IT for personal change, especially with regard to the nonindustrialized world. Since nonindustrialized nations are poised to leapfrog more industrialized countries in their use of mobile technologies, mobile devices are likely to be an important IT vehicle for personal change around the world. As Elaine Huang and Khai Truong (2008) discuss, "Designing phone interfaces and services that help users to engage in sustainable practices with low effort and make informed decisions are clear avenues for which user research and experience design knowledge are critical" (p. 323).

The UN/Vodaphone report *Wireless Technology for Social Change: Trends in Mobile Use by NGOs* (Kinkade & Verclas, 2008) describes a range of mobile projects with positive environmental outcomes. For example, researchers at the University of California, Berkeley and Intel enabled environmental monitoring of air quality in Accra, Ghana, by putting pollution sensors and GPS devices on taxi cabs (Honicky, Brewer, Paulos, & White, 2008). Another project reduced the number of human-elephant conflicts by providing community members and game rangers with "Push to Talk" devices that helped them monitor fences and improve communication in the Laikipia District of Kenya (Kinkade & Verclas, 2008).

Mobile tools and techniques created for environmental information, analysis, and action in the industrialized nations may be used or adapted for use in nonindustrialized nations, and vice versa. Across both industrialized and nonindustrialized countries, mobile IT is likely to have a growing impact on environmental issues over the years to come.

Example Project: Trackulous

A project called Trackulous, which I began building in 2007, provides a useful case study of some of the ways in which IT can help people enact behavioral change.[4] The project is based on the premise that corporations and governments have powerful tools for tracking information to accomplish their goals, but that everyday people lack the resources to work effectively with their own information. This lack of support limits many people's ability to live sustainably, since managing their environmental impact requires accurate and timely information about their own lives. Trackulous is an online system that provides a suite of Web-based tools to help people track, analyze, and share their own personal information—including both environmental and nonenvironmentally related topics—in ways that they themselves find useful. A prototype of the site has had over eighty six thousand hits during twenty-one months of operation. The site's broad goal is to offer a way for people to understand their own personal information in the context of data collected by other members of this online community. By allowing people to work with the vast bodies of information that are important to them, Trackulous can aid them in living well-informed and sustainable lives.

A Well-informed Life

This project was inspired by a realization from my own life. Several years ago I began filling out an online carbon footprint calculator (Nature Conservancy, 2009). When I was about halfway done, the questionnaire asked about how much I fly. I realized that although I fly a great deal to attend conferences, visit NSF, and so on, I did not have an accurate estimate of how much I fly. Not knowing the real answer, I guessed: I fly about once a month or maybe a little more, with most of the trips being long flights. On looking at my results, I saw that, not counting my air travel, my footprint was slightly less than twice the world average and about one-third the U.S. average (see figure 7.2a); including air travel, though, it increased to be slightly greater than the U.S. average and more than five times the world average (see figure 7.2b).

It became clear that, although I ride a bicycle to work, drive a hybrid, and do not eat meat, air travel was a major contribution to pollution (more than two-thirds of my total footprint), and yet I did not have accurate information on this aspect of my life. On further consideration, I came to realize that there were dozens of statistics about

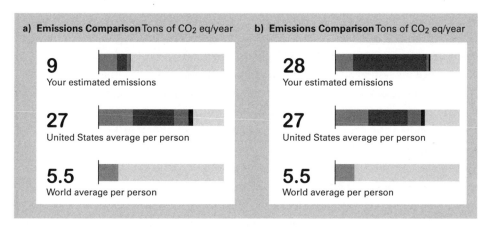

Figure 7.2
My carbon footprint, without air travel (a) and with air travel (b). From http://www.nature.org/initiatives/climatechange/calculator. © Nature Conservancy (http://www.nature.org). Reproduced with permission.

my life that I did not collect and hence did not have access to reliably. On still further consideration, I realized that, as an informatics professor and reasonably competent programmer, I might be able to make a system that could help address this problem.

In addition to the simple lack of data from my own life, I realized too that the world is full of interdependencies that affect the environmental impact of any given decision. For example, when one buys a house, one may not think about the long-term environmental impact of one's ensuing commute, the amount of water the yard needs to stay alive, or whether the township collects recycling. Since the human learning mechanism evolved primarily to predict short-term phenomena such as events that cause pain, the presence of danger in a certain location, or the quality of a food source, people are not inherently good at understanding long-term indirectly connected phenomena, such as whether smoking causes cancer, obesity gives one bad knees, or driving cars too much causes global climatic disruption. Providing people with information about the connections between disparate parts of their lives, especially those that might not initially seem connected, could help them choose behaviors that have more positive sustainability outcomes than information focused only on narrow aspects of their lives.

The core goal of this project is to make it as easy as possible for people to track, analyze, and share their own personal information. By doing so, this project seeks to enable people to live a well-informed life—one in which their decisions are based on accurate and current information. While in many cases, people's actions are necessitated by the cultural and institutional contexts in which they are embedded (e.g., it would be difficult for me to get tenure as a professor without traveling regularly to conferences), there is often a degree of leeway through which people can determine their own environmental impacts. For example, I presented at the 2008 Hawaii International Conference on System Sciences conference via videoconference, rather than in person (despite some reluctance to miss a conference held in Hawaii). Additionally, by taking action in support of one's ideals, one can help shift society's priorities so that institutional support arises for environmentally sustainable lifestyles. As long as humans are living in an unsustainable way (and there is a long way to go to achieve sustainability), Trackulous and systems like it can help people understand their lives and take conscious action to lessen their impact on the world around them.

To people visiting the site, Trackulous might not initially seem to be a Green IT project. The system was initially conceived of as such, but the capability of IT systems to scale made it easy to extend the site to cover many other potential aspects of life. The realization that a whole-lifestyle tracking site might be more effective at helping people track their environmental footprint than a site solely oriented to environmental issues has been one of the organizing principles of the project since early in the development process. Assisting people in living data-backed lives, enhancing their intuition with well-grounded information, is the project's key purpose.

The remainder of this chapter provides a description of a functional prototype, results from a beta test of this prototype, and the design of a revised version currently under development. By presenting the design process of the Trackulous project, this chapter offers insight into how IT can help people enact behavioral change.

Lower the Bar for Participation

The Trackulous system seeks to facilitate the tracking of various phenomena by providing technological and social support for several aspects of the process. By helping people track different phenomena, the system can facilitate behavioral change, build communities, and detect the emergence of trends. Groups of people and other entities are encouraged to perform distributed tracking and analysis, thereby enabling various crowd effects to occur.

In the past, the problem of keeping track of things such as one's weight were solved using pieces of graph paper, Microsoft Excel, or other tools for storing data. The problem of sharing one's results was addressed via word of mouth, special interest groups (e.g., Weight Watchers), or other organizations. The disadvantage was that the lack of a uniform mechanism for tracking and sharing data meant that solutions were created in an ad hoc fashion by each person, and/or for each different phenomenon that might be tracked, and therefore it was difficult to track and share these data. The difficulty of these processes, while appearing mild in any single case, is sufficiently great when seen across all possible users (people, organizations, etc.) and phenomena that tracking and sharing occurs relatively rarely.

By enabling tracking and sharing to be done with a much greater degree of ease, a quantitative increase in efficiency translates into a qualitative transformation in the way people, organizations, and other entities can pursue and understand their activities. The unification of tools for tracking and sharing data and information, and the specific innovations described above, provide a substantial advantage over existing mechanisms.

Implementation

The system uses a central database of all the things being tracked ("trackables") to enhance the value of each of the trackables, and of the system as a whole, to users and other constituents. This database is made accessible for uploading data as well as browsing and analyzing data through an online Web site. Taken together, the database and system for accessing it lowers the overhead of tracking something. Through this process, the system enables people to track things more readily, track them more consistently, and decide to become aware of things they might not otherwise have tracked. The use of a central computational framework for tracking also opens the possibility for fully automated tracking through devices that upload data to the system's database. The broad goal is to help people start tracking all sorts of events and information from their lives with ease, so they have the data if they want to use it to enhance their own understanding later.

In the currently operational system, there are several key data structures:

• *Users:* Stores user name, password, gender, zip code, image, and so on.

• *Trackable types:* Name, description, and data type of a particular group of trackables.

• *Data types:* Name of a given data type that may be shared across trackable types and therefore allow correlations among different trackable types. Could also include relations with other data types (e.g, one hour equals sixty minutes).

• *Trackables:* A collection of samples for a particular user and a particular trackable type. Also contains a visibility field that determines who is able to view the contents of that trackable and an affiliated image.

• *Samples:* A recording of a datum—for example, the number of hours of sleep a given user had on a certain night. Also contains affiliated image and text field.

• *Friends:* A connection between two users that grants additional reciprocal access privileges. Could be extended to enable larger groupings, multiple levels of friends, nested groups of friends, and so forth.

• *Messages:*   Communications between two or more users; two primary forms include private messages and public comments, but various other forms of messages could also be included. Users have the option to have messages forwarded to other media (currently email, but could be extended to other formats).

• *Reminders:*   Communications from the system to the user sent via other media (in particular, email, but could be extended to instant messaging or other formats) to facilitate periodic updating of database.

• *Events:*   Every time something happens in the system (e.g., user inputs data or user accesses a page), an "event" archives an assortment of information about the transaction—time, user name, IP address, details of host computer, and so on.

Track   The list of things that could be tracked is large. Anything that might be reducible to a digital form could be tracked—numbers, dates, text, images, video, and so on. For example, people might track personal habits such as weight, hours of sleep, pages printed, gas mileage, or details of social interactions. They could keep track of community habits, including the number of people in attendance at meetings or contributions to a central resource (shared car, communal account of money, mobile phone minutes, etc.). They could also track environmental phenomena such as the soil moisture at the base of each tree in a forest, the air quality in various locations, or the chemical composition of the water in their kitchen sink.

There are several different ways in which the system might acquire data as well. The most direct pathway is via a human entering it. For other information, though, it may be possible for a computational system to upload it (e.g., a printer that archives the number of pages printed each day). Additionally, an external database could be accessed (e.g., retrieving the number of calories in certain food) to combine with other information sources. Finally, new data could be generated through analysis of existing data (e.g., twenty-four hours minus hours of sleep per day minus hours of work per day equals hours of leisure per day).

Analyze   The Web site currently provides a number of ways to enhance the tracking process. It offers automatic recording of the time of a data sample; the ability to annotate samples with text, images and so forth; graphing capabilities, such as a line graph of change over time; regular reminders, such as emails sent once a day, week, month, or year; and message passing between users. The site serves as a central location for many different kinds of tracking and reduces the overhead of tracking something new, thereby enabling users to adopt a mind-set in which tracking new phenomena seems easy. Also, because of the social networking aspect of the site, Trackulous effectively offers suggestions about things to track based on what other members of the community are tracking. Beyond what has already been implemented, features such as the ability to tag specific trackables or samples could be readily added.

The existence of a central system where one person can track a variety of trackables and where multiple people archive many different trackables enhances the utility of the system in several ways. For instance, it allows users to see what other people are tracking, for inspiration and ideas. Novelty and frequency are two important aspects of suggesting potential trackables—seeing what the most people are tracking, and what people have been tracking recently—but other means of displaying potential trackables can be used as well—alphabetic, random, funniest, and so on. The site allows users to see how other people are doing, for inspiration and/or guilt regarding their own efforts. It permits users to see community averages, for a broader understanding of the social context in which they are enacting change. It provides a way for other people to see how an individual user is doing, thereby enabling social support. Finally, it encourages a feeling of "good citizenship" by contributing to community averages. Just as participating in a blood drive by the Red Cross makes a person feel that they have done something good by donating their blood, people uploading to Trackulous can see it as a contribution of data to their community.

The system can encourage behavioral change in a number of ways. First, it can alter behavior to continue an upward or downward trend in personal statistics (e.g., weight loss or miles driven). Second, it can shift behavior to keep in step with or outdo another individual, friend, or group. It can also change behavior to match or exceed a community average, or a community maximum or minimum. Finally, it can help the user strive to be represented on a community-generated or data-driven list of high or low scorers, or receive other marks of distinction.

The data stored in the various trackables, users, and other data structures provide opportunities to enhance the functionality of the system through data mining. For example, the site could calculate the community averages of various user characteristics. The community in question can be the entire user base or could be restricted based on a variety of other factors (age of user, zip code, gender, what else they are tracking, etc.). These averages could also have outlier values identified and removed, to decrease the impact of people entering random values. Since there is the possibility of inaccurate data, either intentionally or accidentally, the ability to discard high and low outliers and calculate viable averages from the remainder could improve accuracy. The site could also calculate maximum and minimum values for a community, trends over time, correlations between trackables, or correlations between trackables and user attributes. While these capabilities have not been implemented in the current system, they offer some insight into future directions for systems of this kind.

Share Trackulous facilitates the selection of trackables by collecting and sorting the phenomena being tracked by the whole community, and presenting them as suggestions to users so that they may track them also. This functionality makes people more likely to track more things. Second, it causes people to track phenomena using the same trackable name and data type, which then allows for comparison and data mining across multiple different users.

These same benefits also extend to the selection of data types for new trackable types. By helping people select the appropriate unit of measure, the site streamlines the creation of new trackable types. And by standardizing the units selected across multiple trackables, the system allows for further examination of the data by concretizing correspondences across different trackables (e.g., pounds versus kilograms).

The site's "friends" system harnesses the power of social support to enhance its functionality as well. For example, the ability for a user to plot a trackable on the same graph as the same trackable being tracked by a friend enables prompt comparisons without having to contact the friend explicitly for the data.

The site also allows people to form friendships with people whom they have never met but for whom they have merely seen what those people are tracking. This enables people to find not only additional social partners but also social support for their tracking activities (and corresponding life changes) without having to involve existing real-world social partners. Helping people find "change buddies" and coordinate their efforts to alter their lives can provide a substantial value to all involved.

Besides offering additional value to the user, the fact that the system connects different trackables, samples, and users via the trackable type and data type abstractions means that the collected data can be examined more effectively, and thereby create extra value for academics or other researchers who might benefit from this information.

Privacy
There are several privacy issues that go along with a project of this kind. Creating a centralized database with information on many people's personal habits is potentially problematic. The database could be hacked, allowing unauthorized individuals to access it. It could enable data mining, which, even if the results benefited the users, could be disconcerting to those users. Various other more subtle problems could occur if users did not realize that others could see the statistics they are tracking. The development team sought to address these issues by including password protection on users' log-

ins, keeping up to date with server software, and providing users with a set of controls to specify privacy levels (e.g., "Public," "Friends," "Anonymous," and "Private") with which they are comfortable on a trackable-by-trackable basis. Despite steps that have been taken in this project to deal with privacy concerns, the juxtaposition between collecting and analyzing large amounts of data, on the one hand, and enabling people to preserve their own privacy, on the other, offers a key conundrum for environmental IT tools.

Usage

Trackulous has been in continuous public operation from October 8, 2007, until the present (July 28, 2009). During this time, the site has received 86,937 hits, including 28,291 from logged-in users, and 58,646 from users who were not logged in. These hits came from 6,031 unique IP addresses. A total of 553 users registered for the site—a precondition of tracking their own data. These users created 720 trackables of 223 different types (e.g., "Hours Worked" and "Time Studying") and 90 data types (e.g., "Kilograms" and "Hours"). They recorded 3,391 samples of data for these trackables (e.g., the user's weight at a given time).

Most users of the site tracked a relatively small number of trackables, with 88 percent tracking 3 items or fewer; however, individual users maintained as many as 14 different trackables. Four users had more than a thousand hits on the site, demonstrating a certain amount of extended interest in the service provided by the site. Users of the site appear to be particularly interested in the graphing functionality, making 2,213 graphs (1,147 by logged-in users, and 1,066 by non-logged-in ones).

Trackables varied across a wide range of topics. A sampling of trackables includes "Stomach Pain," "Fish Caught," "Miles Ridden," "Songs Downloaded," "Butterflies in My Yard," "Email, Email, Darn Email," "Times I Checked Facebook," "401(k) Value," "Trash I Produced Today," and "Gallons of Gas Filled." The most tracked topics were "Exercise," "Weight," "Hours of Sleep," "How Am I Feeling," "How Often I Eat," "Food", and "Environmental Footprint," suggesting that personal health and fitness are key motivators for participation.

Looking at specific trackables, there is substantial variation in the number of samples. Most trackables have a small number of samples, indicative that the user made a trackable, tried the system briefly, and never came back. A number of users, though, apparently found value in the system, at least enough to come back repeatedly. One user recorded 236 samples for the same trackable over a thirteen-month period, an average frequency slightly higher than once every other day for more than a year. While many people who interacted with the system did not return regularly to the site, some people appeared to find Trackulous to be a useful tool for managing their lives.

Discussion

While Trackulous was designed initially with an environmental focus (with a green color scheme, a hummingbird logo, several initial trackables with environmental goals, and various other minor design elements), it quickly became apparent that people tended to use it more for personal health and life maintenance than for environmental ends. In fact, the two most viewed trackables were "Fights with Boyfriend" and "Everything Rocky Destroyed" (Rocky being a user's dog).

There are several main uses that this system may have in the future, specifically:

• *Personal improvement:* The system can be used to help people keep track of data about their own lives; share this information with their friends, family, or other associates; and enact behavioral change. Organizations such as Weight Watchers and Alcoholics Anonymous have thrived in this area.

• *Environmental action:* There is a growing effort to reduce one's environmental impact. By keeping track of one's own resource use, waste production, and so forth, one can potentially improve one's efficacy in this area.

• *Online dating:* People often like others who are interested in the same topics or dealing with the same concerns. By allowing people to form friendships through the system, the power of the system could translate into a way of starting real-world friendships and romantic relationships. Match.com, eHarmony, Yahoo Personals, and other organizations have been successful in this area; Trackulous could potentially do a better job.

• *Social networking:* With the rise of Facebook and MySpace, social networking has become hugely popular. Tracking different phenomena could become an important part of these existing systems (in fact, Trackulous already integrates with Facebook). Additionally, the system could become a new way of appealing to people who might not be interested in the current characteristics of social networking.

Nevertheless, while this usage was initially less satisfying than if the users had focused their tracking on environmental topics, it potentially points to a broader theme. Rather than seeking to get people to track just the environmental aspects of their lives, it may be more productive to encourage people to live more data-driven lives in general. This concept was what ultimately led to the well-informed life notion. Once people get in the habit of tracking things about their life—whether it concerns exercise, work done, or the daily goings-on of their pet—they will likely realize that they can track all sorts of different topics, some of which will be environmentally relevant. Additionally, if they encounter an environmental issue and want to analyze their own life in regard to it, they will now have the tools to do so effectively.

Broad Goal

A site such as Trackulous may not at first seem transformative; people have been graphing their weight and other important personal statistics for many years. However, enabling people to take on a lifestyle in which they habitually track their data is a major shift in the way people live. It is a transformation similar in nature to (albeit perhaps less significant than) the move that happened several hundred years ago when personal timepieces—watches—became common. Prior to watches, two people could not carefully coordinate their meeting time; they could schedule their lives with quite limited precision. With the advent of watches, however, two people could arrange to meet at, say, 10:20 p.m., and each could arrive at that time with a reasonable certainty that the other party would be there as well. The move facilitated by Trackulous and other similar sites, through which people begin to live a well-informed life, could usher in a transformation of a similar quality, if not degree, to the transformation regarding time. No longer will it be socially acceptable to dissemble when asked what one's gas mileage is; if everyone uses trackers ubiquitously, one will be expected to know such figures. While privacy issues present important challenges in such universal tracking, human societies have many effective social and technological techniques for ensuring privacy in similar systems. If people habitually archive and analyze their own personal data, they can begin to act with much greater precision to change their lives to match their ideals.

EHCC Perspective

IT systems for enabling personal change help contribute to the EHCC perspective in a number of ways. First, by giving people greater control over their future behavior, they help people extend their own time horizons. People with a greater sense of efficacy in determining their own future tend to take a longer time horizon when thinking about their future (Bembenutty & Chen, 2005). Providing people with the tools to change their own behavior and the world around them can aid in addressing the unfortunately common "There's nothing I can do, so it doesn't matter" argument against long-term thinking and action.

By showing people that it is possible to enact personal changes, Green IT systems can keep the focus on human change as a critical factor in achieving environmental sustainability. Researchers studying how people learn found that individuals who see intelligence as something they can practice and get better at (like strength) tend to do better than those who think it is an innate ability (like height) (Dweck, 2006). Enabling people to realize that they can be part of the solution to the world's long-term challenges is a large part of spreading an extended human-centered perspective.

Tools for enabling personal change can also provide value by allowing people to integrate their own behavioral decisions with command-and-control decisions made at broader institutional scales. Since the operation of many human institutions and infrastructures involves IT tools for optimizing efficiency and other factors, the spread of tools for personal change may permit people to "plug in" to these broader systems, thereby enhancing both their own efficacy and contributing to the effective operation of the collective.

This chapter has offered a range of Green IT efforts that enable individual behavioral change and has presented an in-depth discussion of Trackulous as a case study in this area. Green IT efforts are having an impact around the world with people from many different cultures. There are systems that help people find sustainable resources, understand the impacts of their own actions, and hack their lifestyles to make them more sustainable. A variety of effects are arising from these systems, including direct reductions of environmental impact, improvements in quality of life, feelings of association with environmental ideologies, and the encouragement of social partners to take action as well. Taken together, the effects of current Green IT systems for facilitating personal change are substantial; the future potential of this field is even greater. Tools for aiding personal change lie at one end of a continuum of IT interventions arranged around broadening social complexity. The next chapter will explore ways in which Green IT also helps build communities and support collective action around environmental themes. Efforts at the personal, communal, and institutional levels will need to be undertaken and integrated together for a viable large-scale response to current environmental concerns to occur.

8　Green IT and Collective Action

I will completely stop using disposable cups for tea/coffee at work but only if 100 other people worldwide will do the same.

—Julia

Deadline to sign up by: July 31, 2005; 106 people signed up (6 over the target).

I will replace all of the traditional light bulbs in my home with low energy versions but only if 60 other people will do the same—and ideally persuade more people to do it too!

—Steve Jalim

Deadline to sign up by: December 31, 2006; 62 people signed up (2 over the target).

I will for 3 months save water by putting a bucket in the shower while it is warming up, limiting my showers to 4 minutes and using the bucket to wash my hair or water the garden but only if 20 other people will try their very best to do the same.

—Lisa, average person on a humble mission

Deadline to sign up by: January 1, 2007; 20 people signed up.

I will give £10 towards a tree planting project in Kenya but only if 200 other people will do the same.

—Chris Anderson

Deadline to sign up by: March 1, 2007; 211 people signed up (11 over the target).

I will Go Vegetarian between 25th–29th June 2008 but only if 15 other meat eaters will do the same.

—Green Festival Man

Deadline to sign up by: June 25, 2008; 16 people signed up (1 over the target).

The above quotes are taken from a Web site called PledgeBank (pledgebank.com, 2008) designed to help people find like-minded individuals and support their collective action. The goal is to overcome some of the inertia and loneliness that often thwarts potential action by environmentally minded individuals.

The environmental issues facing the planet are daunting. With 6.7 billion people on Earth, the impact of one person can often seem insignificant. Taking action on one's own may seem futile—exhausting without being rewarding. Especially when one considers how much of one's environmental impact is prescribed by the infrastructure and societal norms in which one is embedded (clothing choice, lifestyle, etc.), the margin of individual impact seems small.

IT systems such as PledgeBank, however, are beginning to enable collective action in ways previously impossible. Communities frequently act across broader scales of time, space, and complexity than individuals, and may have broader impacts than most individual efforts. By increasing connectivity and communication between like-minded people, and providing support for their collective action, IT can enable synergies that make the impact of individual contributions more apparent, and thus incentivizes further participation.

This chapter explores a range of ways in which IT supports communities and collective action, and uses a particular project, called GreenScanner, to examine in depth one way that people may be enabled to work together to live more sustainably.

Audiences

Effective community-based Green IT has a broad potential reach across a range of audiences. With hundreds of millions of desktop computers and billions of mobile phones, the technological infrastructure is in place for massive and rapid mobilization of human action.

Perhaps the most apparent form of community Green IT involves systems that are being used in the industrialized world to help people work together to reduce their environmental impact. By building on existing social networks (both online and off-line), and creating new connections between people and organizations, these systems can strengthen communities while they encourage people to act together.

In addition, there are at least two main ways in which community Green IT reaches people in the nonindustrialized world. First, systems can affect them directly, facilitating cooperation through mobile devices as well as desktop computers. While people in the nonindustrialized world may be reluctant to compromise their quality of life for environmental ends, there are many ways that communities working together can achieve a higher quality of life by engaging in environmentally preferable behaviors, rather than despite them. New forms of mobile collaboration, resource sharing, and other community action can simultaneously benefit individuals' short-term happiness and enhance the global ecosystem's long-term survival.

Second, Green IT enables connections across the world, providing mechanisms through which people with different resources (such as money, time, and access to critical ecosystems) can collaborate effectively to leverage their respective strengths. Global-scale community action—for example, where people in industrialized nations offer financial support for on-the-ground action in places where it can have the greatest impact—could be powerful. (Note that this strategy is not intended to suggest that people in the industrialized world can buy their way out of practical action beyond mere financial contributions. All people will likely need to revise their behavior in some way, particularly those individuals with the greatest environmental footprints. Nevertheless, financial support for action in other places around the world is certainly a valuable contribution.)

While the industrialized world currently has a much greater environmental footprint per person, the nonindustrialized world may soon approach that same level of impact. Instead, now is the time to establish community-based environmental innovations that enable the nonindustrialized nations to leapfrog the environmental practices in place in the industrialized world on their way to a high quality of life.

Impact

IT can help communities respond in a number of different ways: building social networks, spreading ideas for change, encouraging specialization, and sustaining group action.

Build Networks

Social networking systems such as Facebook and MySpace are causing human social networks to take on new dimensions, letting people connect with off-line friends in new ways, enabling them to find new friends online, and helping them discover new social groups. The design of social software encourages the formation of different kinds of relationships and supports a range of collective activities.

Through online social networks, people may stay in touch more effectively with people whom they have met off-line, and also find others who share their interests or ideologies (Ito et al., 2009). The latter effect is particularly important for enabling environmental action; finding groups of like-minded individuals can provide social support and opportunities for collaboration that would be harder in off-line contexts.

Beyond the goal of finding friendships and collaborators, people can use social networks to look for romantic relationships and life partnerships that support their sustainable ideals. Humans have an uncanny ability to co-opt the functionality of networked computational systems to cause them to serve as dating sites. In an early example of this phenomenon from 1995, Firefly, a site that helped introduce people to new kinds of music, rapidly gained use as a tool for finding dates who liked the same kind of music. A large number of Green online dating systems have sprung up, with names like planetearthsingles.com, greensingles.com, and green-passions.com. In a testament to the popularity of this genre, National Public Radio discussed a Green speed-dating event to help environmentally minded people meet others who share their interests (Bergman, 2008).

Enabling people to connect, whether romantically or platonically, can support many other kinds of beneficial interactions.

Spread Ideas

Another valuable mechanism by which Green IT can enable community-based environmental efforts is by encouraging the spread of ideas. Human communities act as a kind of distributed learning engine—each person goes out and interacts with the world, learns from that experience, and then shares the results with social partners via communication acts. This process is beneficial because it means that people can learn from indirect experience without having to undergo the danger or other challenges associated with direct experience. Distributed learning enabled by communication is one of the most central attributes that has led to humans' global success; perhaps in an IT format it can also help address the environmental concerns created by that success.

At the conference Text and Image: From Book History to "The Book Is History," held at the University of California, Irvine in February 2007, Ramesh Jain offered three key elements that allow humans to learn from each others' experience. The first is the ability to preserve that experience. This preservation may occur by means of human memory or more permanent archives such as books or digital content. The second is the ability to propagate the experience among different individuals—for example, via spoken language or media for the experience preservation mentioned above. The third element involves presenting the experience in a specialized way—for instance, via summarization, visualization, or other presentation techniques.

Green IT can support all three of these elements. Digital archives of text, images, sounds, and video are tools that are well suited for preserving experiences. Computer-mediated communication is well studied, and resonates with the need to propagate experience among different individuals. Finally, computer tools for interactive and noninteractive visualization can enable people to present their own experiences or the collected experiences of others in a specialized way.

Ideas can spread through communities in a number of ways. Peer-to-peer sharing provides one mechanism, as friends, collaborators, and social partners exchange ideas.

This process is called "horizontal transmission." Projects such as Climate Navigator enable this kind of idea dissemination, seeking to "build a virtual community of policymakers, businesspeople, and scientists," and contributing to environmental understanding and action by "aggregating their knowledge, distilling their insights, and publicizing their debates" (CITRIS, 2009).

Another mechanism, called "vertical transmission," involves learning between generations, especially between parents and children. The most well-known form of vertical transmission involves the upbringing process, as parents help their children understand the world. However, there is an interesting form of learning that mirrors this process—when parents learn from their children. Since children around the world spend a great deal of time in schools, where they are exposed to ideas beyond the scope of what their parents know, they often bring ideas back to their parents. Therefore, schools provide a valuable mechanism for spreading environmental concepts not just to children in classes but to their parents and other community members as well.

Parent-child relationships are a powerful part of environmental efforts. In addition to learning from their children, parents appear to be more likely to adhere to their own ideology when acting in front of their children. Sometimes it can be challenging to act in accord with one's own beliefs when there are so many other competing issues. For example, I sometimes fail to recycle a can if there is not a recycling bin nearby. If I were with a child, however, I would be more likely to make the effort to track down a bin. Since many people believe in the value of sustainability but have a hard time following through on those topics, providing contexts that encourage people to live in accord with their ideals can be of critical assistance in helping them live more sustainably.

A variety of online resources support the spread of environmental ideas. Pledgebank, the source of the quotes that opened this chapter, allows people to see what ideas others are having about how to live greener and then participate if they agree. Sites such as TreeHugger and Worldchanging provide an assortment of suggestions about how to live more sustainably, produced by both experts and community members. Wikipedia, too, with numerous articles about many different environmental topics, is a valuable resource for spreading environmental awareness and information.

The spread of ideas has value beyond supplying people with information and collaborators. In addition, it offers a degree of exposure that enables good ideas to achieve social acceptability or even cachet, and therefore to have a much broader impact. This "cool factor" greatly benefited the makers of the Toyota Prius and other hybrid vehicles, for example. By having movie stars and other celebrities drive and be seen in hybrids, it became apparent that hybrids are *the* car for environmentally concerned individuals.

Encourage Specialization

One of the great challenges behind environmental action lies in knowing what to do. All of us have multiple responsibilities in our own life to contend with—such as jobs, school, parents, children, and friends—and do not always have the time or energy to spend figuring out what to do. If given a clear path ("buy a hybrid car," "eat organic," or "reduce, reuse, recycle"), people may be willing to participate. However, given a relatively small amount of time to figure out what to do, people often do not trust that they have sufficient understanding of the issues to make an educated decision. Thus, they tend to prefer to off-load this responsibility onto other organizations—such as the U.S. government, the Sierra Club, and so on—to make the decisions for them. In some of these cases, the organization does not necessarily have environmental goals as its first priority.

In order to enable people to go about their lives in a sustainable way, it will be necessary for someone to attend to these issues for them. This off-loading of decision-making responsibility entails several factors:

• *Delegation:*   An individual must somehow empower another individual or organization to make decisions on their behalf. The most pragmatic mechanism for this delegation is to pay them, so that they do not need to have some other job.

• *Communication:*   Once another entity is "doing the thinking" on another person's behalf, it is necessary for that entity to communicate the appropriate actions and possibly the reasons for those actions to the person. Because of the rapid rate of change of environmental issues, this communication might benefit from embracing new communication technologies. A monthly newsletter is not necessarily sufficient; rather, instantly available current information via a mobile phone, the Web, or some other technology would be better (as long as the technology itself is not contributing too greatly to environmental problems).

• *Trust:*   Individuals need to believe that the entities to which they have delegated environmental decision making have the same goals and desires as the individuals themselves. Presumably this is usually the case when individuals make the decision to off-load responsibility. In addition, though, it is necessary for this trust to continue throughout the relationship.

• *Adherence:*   Given that individuals are receiving information that they trust in a timely fashion, it is then necessary that the individuals act on this information.

Therefore, perhaps it is time for another profession in society—a personal environmental consultant. Just as there are hairdressers, doctors, lawyers, therapists, priests, and accountants, each serving some unique role in our lives, one's environmental consultant could serve the role of providing people with up-to-date environmental information tailored to the particular circumstances of their lives. This person would need to know many of the details of one's life in order to comment meaningfully on environmental issues relating to work, home, family, health, and numerous other concerns. The person would need to take a holistic view of one's daily experience in order to understand which issues would make the most difference. There would certainly be privacy concerns with such a person; but just as therapists, doctors, lawyers, and priests have ethical standards, so too could environmental consultants.

Creating sociotechnical support for this new profession could help it become a reality. This support would be a two-way communication mechanism by which clients inform their consultants about what is going on in their lives (either directly, by active communication, or indirectly, by means of some automated process), and consultants can then offer suggestions regarding behavioral choices to take. Enabling these consultants to access information (e.g., from buildings, vehicles, or corporations) and take action (e.g., setting parameters on a smart power grid) directly could streamline the end-user experience (albeit accompanied by a variety of privacy concerns and other risks). In addition, a whole suite of sociotechnical tools could be used by groups of environmental consultants working together to share information and thereby arrive at more effective decisions.

Enabling some people to devote full time to figuring out the best responses to various environmental concerns, and having other people trust that their suggestions are the best available given the current information and should therefore be followed, can support well-informed action on a large scale. While it might not be worth experts' time to investigate some small issue in order to inform one individual's decision making, if they knew that ten million people might take their advice, it could be much more worthwhile to research it. Green IT can support this effort by helping experts discover what issues will provide them with the greatest impact, supplying them with research tools, and assisting them to convey the results of their research to the largest number of people in a way that those people will understand.

Encouraging specialization at various levels—within local communities or internationally—can be a powerful way to enable people to utilize their strengths most effectively, and cause groups to achieve the greatest usefulness from their actions.

Sustain Group Action

Perhaps the most direct way that Green IT can enhance community responses to environmental concerns is by supporting group action. This action can occur at a local level or more broadly, encouraging corporations and governments to pursue environmental goals.

The World Naked Bike Ride, mentioned earlier, is indicative of the ways in which IT is facilitating action on a broad scale. With the rise of the World Wide Web, rapid global action is becoming more feasible. Efforts of this magnitude previously would have required slow or expensive forms of communication and coordination across great amounts of space, and been vastly complex. Now, such action can occur much more rapidly, the complexity to the individual has been reduced (aside from the complexity involved in the existence of technological and social infrastructure for global communications), and the accessibility of participation in such action has been increased. In this undertaking, individuals with a range of motivations (e.g., supporting cyclist safety, encouraging renewable energy, and engaging in good-natured exhibitionism) found each other via the Internet, formed strategic alliances, and coordinated a collective effort to make their voices heard. IT effectively shrank the world both physically and socially, made possible a new form of community involvement, and supported a novel effort to encourage sustainable living.

At a local level, Green IT can provide tools for coordinating groups in terms of synchronization, cooperation, and the division of labor. Computational communication technologies such as text messaging have been rapidly co-opted by protesters and organizers seeking to gather large groups together quickly, and include others who cannot be present physically, such as happened during the World Trade Organization protests in Seattle (Sullivan, 1999). Various tasks that require multiple people are often arranged via online systems, such as the Surfrider Foundation's Beach Cleanups. Communication via IT can also help groups to split up their work effectively and provide real-time feedback on the progress of various subgroups, thereby avoiding redundant effort and permitting more rapid adaptation to changing circumstances.

Community action is frequently directed at encouraging corporations to engage in more environmentally aware practices. As an example, a project called Carrotmob "organizes consumers to make purchases that give financial rewards to those businesses who agree to make socially beneficial choices" (carrotmob.org, 2008). Essentially, the organizer talked to all the local markets in his neighborhood, saying that he would bring a group of people to shop at a specified time to whichever store pledged to use the highest percentage of their income during that time for the environmentally sound retrofitting of their store. The winning store pledged 22 percent. The organizer put out word of this effort through his social network, and at the appointed time, several hundred people arrived, more than doubling the store's expected sales for the day.

Conversely, people often organize boycotts of companies based on their environmental policies. Environment-based boycotts have been waged against Exxon (Overby,

2005), Mitsubishi (Luo, 1995), Royal Caribbean Cruise Lines (Phillips, 2003), and many others.

Some companies' business models rely on environmental consumerism. Zipcar allows people to share automobiles with ease. The Method brand offers Green cleaning products. SmartBike DC is a joint private-public venture to encourage bicycle sharing (Becker, 2008). Many new companies are arising to participate in the new Green economy. All of these efforts are enabled by communities of environmentally minded individuals with strong voices; these voices can be heard more strongly now that IT allows them to speak in unison.

IT also enables community action on a governmental level. The Internet has proved to be a critical aspect of a new style of grassroots campaigning, playing an integral role in several U.S. presidential campaigns, most notably those of Howard Dean in 2004, and Ron Paul and Barack Obama in 2008. A number of environmentally related online systems—for example, the League of Conservation Voters (lcv.org, 2009) and StopGlobalWarming.org (2009)—offer a way to send emails to congresspeople or other governmental officials, or other ways to take action.

Taken together, the environmental impact of IT systems that permit people to meet each other, share ideas, and work together is substantial, and growing.

Example Project: GreenScanner

To provide an in-depth example of an IT project designed to help support community action around environmental themes, the remainder of this chapter describes Green-Scanner.[5] GreenScanner (see figure 8.1) is a system that helps people engage in environmentally preferable purchasing during their everyday consumer transactions. This system has two parts: a dynamic online database of community-generated, environmental-impact product reviews, and a mobile phone application that allows consumers to access these reviews at a point of purchase by photographing a product's Universal Product Code (UPC). The remainder of this chapter explores a prototype of the system that was released publicly in April 2006, and analyzes the results of that deployment. Over its first year of existence, the site received approximately thirty-one thousand hits. During that time, though, only seventy-nine subjective comments were uploaded. The site is no longer under active development; nevertheless, a number of commercial systems are now providing aspects of this service (see, for example, goodguide.com, 2009; greenerChoices.org, 2003; montereybayaquarium.org, 2009). This chapter looks at GreenScanner's usage patterns, and discusses how developer and user experiences with this site can lay the groundwork for future efforts in this growing domain.

Introduction

As the global population increases and the demands for natural resources approach or exceed their limits, a critical issue is the conservation of resources. An important problem in U.S. society is the great deal of waste produced as a result of the manufacture and consumption of consumer products. As discussed previously, the United States produces 230 billion kilograms of municipal solid waste per year, or more than 2 kilograms of trash per person per day (U.S. Environmental Protection Agency, 2008a). A sustainable environment mandates the reduction of consumer waste.

Many consumers have at least some degree of environmental goodwill and the desire to engage in environmentally preferable purchasing (U.S. Environmental Protection Agency, 2006), but do not have access to reliable information in this domain at the time of purchase. In the absence of this information, they may instead choose products based on price, labeling, imperfect memory, or force of habit. To help provide useful information to consumers at the point of purchase, various rating systems have been proposed, such as those relating to the healthiness of products (see, for example, Martin, 2006). Product labels sometimes include statements regarding the environmental impact of products; however, these labels are not ubiquitous, they may be inconsistent, and often their relevance is not clear. Ultimately, consumers still may not know how environmentally sound different products are. Environmental considerations are clearly a priority for federal purchasing (U.S. Environmental Protection Agency, 2006); the goal of the GreenScanner project is to achieve similar ends with large numbers of consumers.

Much of the waste produced in the United States comes from products that have UPCs or similar identifying information on them. The term UPC has been renamed UCC-12, and its functionality subsumed by the broader EAN/UCC-13 standard. For ease of understanding, all of these machine-readable tags are called UPCs here. Of the solid waste mentioned above, 59 billion kilograms come from the "glass," "plastics," and "metals" categories, which frequently have UPCs, or 0.54 kilogram of UPC-labeled waste per person per day (U.S. Environmental Protection Agency, 2008a). Corporations and stores use these markings to identify and track their stocks of goods, and enhance their efficiency and productivity. Individual consumers nevertheless have not traditionally had the ability to use these markings effectively themselves. The GreenScanner project lets shoppers use the UPC system for their own purposes—in this case, to access information about environmental impacts.

To summarize, consumer waste is a major problem, and shoppers frequently do not have sufficiently useful information to make environmentally sound purchasing choices. UPCs help companies to access relevant information about products, but these tags

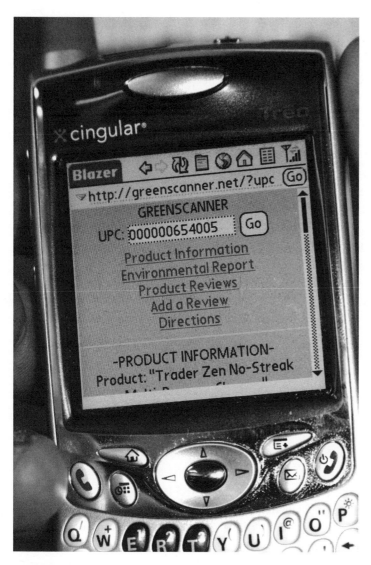

Figure 8.1
GreenScanner provides environmental impact reviews of consumer products via mobile phones.
© 2008 IEEE. Reprinted with permission.

are rarely used by shoppers to access information that is of relevance to them. Green-Scanner helps to solve these problems, coupling mobile phone software for reading UPCs with a database of information about the products' environmental impact. The component technologies are readily available; constructing the system, populating the database, and expanding the community of users have been the goals of this project.

Vision

The following scenario presents an instance of the potential use of GreenScanner. A father and daughter go to the food store to shop together. They walk down the canned vegetable aisle, looking at the different brands. "Do you want peas or corn tonight, kiddo?" the father asks. "Corn!" the daughter replies. They move their cart to the corn section. "Hmm, this can is 99¢, and this other one is $1.09. I wonder what GreenScan-ner has to say about them?" The father gives his daughter his mobile phone. They have made a tradition out of shopping like this—Dad picks two choices, and she gets to pick between them based on the results from GreenScanner. The daughter aims the camera on the phone at the each can of corn in turn and clicks "Scan." The results rapidly ap-pear: for the $1.09 can, GreenScanner displays a score of "6.3," with text that reveals that the corn had been sustainably farmed and packaged in a recyclable can. For the 99¢ can, a score of "5.9" appears, with less favorable reviews. The daughter picks the $1.09 can. The father is glad to spend a dime for his daughter to learn respect for envi-ronmental conservation.

Related Work

The idea of connecting information to products through unique ID tags has a sub-stantial history in corporate contexts. UPC, ISBN, VIN, RFID, and many other pro-tocols have been developed to allow various industries to track their production and inventories. These unique IDs facilitate a wide variety of business processes and make them more efficient. Certain companies primarily use their own proprietary bar code systems for tracking, which makes the integration of their content into comprehensive databases more challenging. Many manufacturers and distributors, however, use stan-dard tagging technologies; it is with these companies' products that GreenScanner is designed to work.

The existence of these standardized ID mechanisms also allows other individuals and groups to exchange other kinds of information about those products. Co-opting these ID processes enables stakeholders whose interests may not be directly aligned with a given corporate entity to construct their own bodies of information about various products. For example, sites such as upcdatabase.com (2009) allow people to access a variety of product information online, in particular discovering what company makes a given product and what other products that company makes.

Accessing these identification systems via mobile devices has been done for several years in grocery and department stores, where handheld UPC scanning "guns" allow employees to identify many different products. Using consumer-owned mobile devices to scan these identifiers was pioneered by the AURA project from Microsoft Research (Brush, Turner, Smith, & Gupta, 2005), which used pocket personal computers to scan UPCs and allow users to view, store, and share metadata and annotations about different products. GreenScanner was the first project to have focused on utilizing this kind of process specifically for environmental ends.

In terms of the connection between consumerism and environmental concerns, the EPA has offered its guidelines for Environmentally Preferable Purchasing (U.S. Environmental Protection Agency, 2006), which provide information for consumers on the environmental impact of a range of different products. While this site has a large amount of worthwhile information, it is not designed to allow consumers to access the information with the necessary speed and convenience to enable it to influence individual purchasing decisions. More recently, commercial efforts such as PriceGrabber.com's (2009) ShopGreen system have also sought to address the needs of environmentally concerned shoppers. Here again, though, the system is not designed for mobile phones. Jennifer Mankoff, Deanna Matthews, Susan Fussell, and Michael Johnson (2007) provide an excellent summary of ways in which social networking sites may be used to foster environmental concern, and describe a proposed project to use these sites to encourage behavioral change.

Many different Web sites allow communities to exchange content in rapid and efficient ways. Urban Dictionary (urbandictionary.com, 2009) is a community-generated database of definitions of slang terms. It allows users to upload definitions of terms, and uses a thumbs-up and thumbs-down mechanism to permit other users to evaluate the quality of those definitions. RateMyProfessors (ratemyprofessors.com, 2009) provides an opportunity for students to write candid, anonymous reviews of their university professors, and rank them by "clarity," "easiness," "helpfulness," and "hotness." Epinions (epinions.com, 2007) allows consumers to evaluate products on a five-star scale, and write reviews of those products. Wikipedia (wikipedia.org, 2009) lets people contribute to community-generated encyclopedia articles. Each of these sites has characteristics that helped to inspire the various versions of GreenScanner, as discussed later in this chapter.

Prototype

A prototype of GreenScanner (see figure 8.2) became public in April 2006. Since it used standard Web technologies, the prototype ran on all Web-enabled mobile phones by accessing a SQL database of consumer products through a PHP Web site. An archived version of the system is available at http://GreenScanner.net.

The database of environmental impact information underlying the GreenScanner system stores several main tables of data. First, it stores product information about all of the consumer items about which information is being collected—product name, size, and company—in an "Items" table. There is also a "Companies" table that includes names for each of the companies that users have added to the system. These tables are populated with information for over eight hundred thousand products, downloaded initially from upcdatabase.com (2009) and expanded by GreenScanner's user community. Environmental impact reports are stored in two tables—one for "Product Reports"

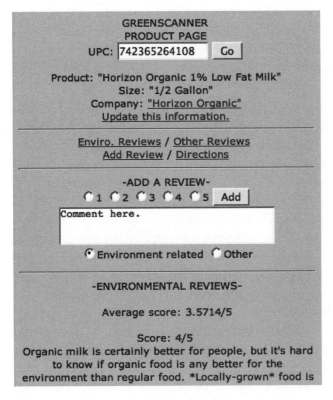

Figure 8.2
GreenScanner product page. © 2008 IEEE. Reprinted with permission.

and one for "Company Reports." Finally, there is a table for "Visits," which keeps track of all of the actions taken by site users.

These tables are manipulated by several Web pages. First, there is an "Index" page, which contains directions for using the site and a text box into which people can type a UPC. When a number is entered, the site confirms that it is a valid UPC and then takes the user to an "Item View" page, which displays the product name, its size, the company that produced it, reviews written by other users, and link to several other pages. The user may enter a review of that product and score it on a scale from one to five. In addition, there is a "Company View" page, which provides a list of all the products by that company and a list of all the comments written by users about that company. Finally, there is an "Update" page where users can make changes to the name of the product, its size, and the name of the company that produced it.

An element of the design of the initial GreenScanner site was the ability to add product and company reviews that were both environmentally related, such as how much waste it produces, and nonenvironmentally related, such as how it tastes. Below the text box into which the review can be entered is a radio button to select between these two choices. These two different groups of reviews are displayed separately on the page, along with the average of the scores that users had given the product.

GreenScanner was designed to be accessed primarily from mobile devices. This focus has led to several design decisions. First, it is formatted to fit on a small screen; all critical information is either visible (e.g., UPC) or linked to (e.g., directions) in the top two inches of the screen. The site also does not include images, video, sound, or any other bandwidth-heavy media, so that it loads as rapidly as possible. Finally, the site was designed to use a standard Web browser, so that compatibility issues could be minimized.

To simplify the process of entering the UPC into the mobile device, the system was designed to use the camera on a given phone to analyze the UPC and input its numeric equivalent into the browser automatically. The initial version of this system was implemented for use on a Palm Treo 650; a freeware application called EAN13Barcode, which uses the Treo's camera, was integrated into the system. In addition, the system could work with a Socket SDIO In-Hand Scan Card, which greatly enhanced the reliability of the UPC scanning. A few months after the public release, a collaboration with Microsoft Research's AURA project allowed Windows Mobile devices to use their cameras to enter UPCs with the GreenScanner system as well. The AURA system required a custom-made macro lens attachment to perform the scanning, available for free from Microsoft Research. All other systems needed to type in the UPC by hand.

Evaluation

The GreenScanner prototype had a number of successes and shortcomings, described here. The methodologies used in assessing the system included analyses of the server logs, scanning process, community-generated content, and external reviews of the GreenScanner system.

Public Reception   A Google search for the term "GreenScanner" produced only 3 hits on April 3, 2006, the day before the site's public release. Within several weeks, the same search revealed more than 15,000 hits. As of June 4, 2007, that search produces 702 hits. Note that due to the structure of sites such as de.licio.us, which cause the same person's tagging of a site to appear multiple times in Google, these numbers are significantly larger than those of distinct human-usages of the relevant term. For example, the June 4, 2007, search located only 57 distinct references to GreenScanner after accounting for the multiplicity of de.licio.us hits.

The rapid rise and then fall of GreenScanner's Web presence probably resulted from the fact that GreenScanner was blogged on a number of prominent sites, including worldchanging.com, grist.org, globeandmail.com, researchbuzz.org, smartmobs.com, ecogeek.org, and supervegan.com. These blogs included comments such as "gloriously simple" (Das, 2006), and "It's cool technology, it's empowering individuals, and it's allowing us to [make] more environmentally informed decisions. . . . [GreenScanner] is the essence of EcoGeek" (Green, 2006). GreenScanner was also a finalist in the Toshiba Green Innovation Awards, an award for environmental technologies in Orange County, California. All of these elements point to a public perception that GreenScanner could be useful in addressing some of the world's environmental issues.

Usage   The GreenScanner site received 31,061 hits from April 4, 2006, through June 5, 2007, or approximately 73 hits per day. This activity started with several weeks of high usage immediately after the university's press release and the ensuing news articles, then dropped off over the next few weeks, and then climbed steadily, albeit slowly, over the following year.

Of the total hits, 19.6 percent were accessing the site's Index page without a UPC, meaning that they were probably visiting for the first time of a given session and had arrived there via a Web link. Approximately 43.7 percent of the hits came through the Product page and included content in the UPC field, meaning that the user had typed something into that field or was accessing the site via a mobile device with an automatic scanning system. The remainder of the hits was split fairly evenly between views of Company pages (18.1 percent) and visits to the Update page (18.6 percent). These visits came from 4,492 distinct IP addresses, or 6.9 page views per IP address.

A total of 2,364 distinct text entries were input by users into the UPC field. Of these, 934 UPCs were found in the product database; the remainder was composed of either UPCs for products that were never added to the database or malformed UPCs (words, mistyped UPCs, etc.). Terms such as "Powerbook," "levis," "green," and "Tide," which people entered in the UPC text field, point to an issue in which people expected certain behavior from the system beyond what it was designed to do.

Four UPCs were publicized broadly in the media: "Horizon Organic 1% Low Fat Milk" (742365264108) and "Poland Spring Water" (075720008513) via a university press release (Fitzenberger, 2006), "Clorox Disinfecting Wipes/Lemon" (044600015941), and "ZBar—Chocolate Brownie" (722252194138) via a Grist.org article (Schults, 2006). Both of these articles were cited by other news articles and blog posts, thereby multiplying their effect. These four products alone had a total of 2,203 hits during the study period. In addition, probably due to the ease of navigation within GreenScanner from one product to the company that produced it and then to a different product by that same company, a large number of other products by these four companies also received a relatively high number of hits. Items by these companies accounted for a total of 20,737 hits, or more than two-thirds of the overall hits for the site. Removing all products by these four companies from this analysis reveals a more accurate picture of long-term site usage, since the goal is to cultivate users who utilize the system in most or all of their purchasing decisions, rather than just visiting once from a news article link. Four products that were used regularly in our research demonstrations were also removed, as were any hits that had a blank UPC (i.e., first visits to the site). What remained could be termed usage by "regular users."

Regular users accounted for 3,973 hits over the study period. There was an average of 9.3 hits per day. A total of 683 valid UPCs were entered at least once. These UPCs ranged from "Oh Naturel Meatless Chick'n Strips" (062000597812) to "Energizer AAA Battery" (039800045638) to "Ez Wider Papers" (074460000016). None of these UPCs had more than 14 hits, and 640 of them had 5 or fewer hits. Multiple hits were usually caused by a single user visiting various pages on the GreenScanner site with the same UPC. The fact that these hits were spread evenly over so many products suggests that there is an abundance of different products in common use within GreenScanner's community.

Ultimately, these data confirm that GreenScanner's users were at least partially utilizing the site as it was intended—to access information about distinct consumer products. In future sites we develop, the team plans to continue collecting similar data to help elucidate the usage more fully.

Reviews   A total of 79 product or company reviews were entered, including 38 "environmental" product reviews, 31 "other" product reviews, 5 "environmental" company reviews, and 5 "other" company reviews. Of these, 31 of the environmental product reviews, 30 of the other product reviews, 4 of the environmental company reviews, and all 5 of the other company reviews were for products or companies not excluded earlier, and thus appear to have been uploaded by regular users. Comparing this total of 70 reviews to the total number of valid UPCs uploaded by regular users, it becomes apparent that over 10 percent of products (70 out of 683) scanned by people were also reviewed by them. While the numbers are fairly low, this result suggests that regular users are relatively willing to provide content to the site.

A number of the environmental reviews contained relevant and reasonably high-quality content. For example:

Twleve [sic] ounce aluminum can packaging is not as environmentally sensitive as 2-liter plastic (PET) bottles. (See http://www.ilea.org/lcas/Tellus.html). Coca-Cola admits they use high fructose corn syrup that was produced with the use of pesticides (see http://www2.coca-cola.com/citizenship/environmental_report2004.pdf). The Coca-Cola company exposes the public to toxins and depletes local public water sources in India (see http://www.ipsnews.net/news.asp?idnews=29971).

Other reviews made environmental claims without documenting the validity of those claims, such as "Nabisco was accused of releasing hazardous chemicals into the environment." Still other environmental reviews sought to encourage behavioral change, rather than supplying factual information: "Paper towels are an uneccesary [sic] waste of paper. Don't buy these, just keep a sponge in your kitchen. You don't need paper towels! They are the lazy person's way, and bad for the environment to boot."

There was also at least one case of likely self-promotion, in which a product review and a company review (below) were submitted in quick succession.

Today there is a growing public concern for reducing damage to our environment and dependence on fossil fuels. GreenFiber's Cocoon insulation responds to this concern by contributing to environmental responsibility. Cocoon has the advantage of locally available raw materials which consists of 85% recycled content. Cocoon requires less embodied energy—the sum of energy required to obtain materials, manufacture and transport a product. Cocoon's responsibility doesn't end with delivery of the product. When installed, Cocoon improves air, thermal and acoustic environment of a home. From raw material through installation, Cocoon builds environmental responsibility.

While self-promotion is not inherently harmful, since individuals and corporations often have a great deal of information about their own operations, it does demonstrate the potential for exploitation of a system of this kind. Users also occasionally added spam reviews and other unhelpful content, such as "Freakin' hippies" or "Seriously, this is good milk. Buy valium and make your lady happy."

Despite the broad assortment of types of reviews uploaded, there did seem to be a facet of the community willing to contribute meaningful content to the GreenScanner database. Nevertheless, the system lacks sufficient environmental impact information for it to be broadly useful as a tool for shopping at this point. Because the system is simply an infrastructure for the exchange of this information, it did not initially have any information beyond the few reviews supplied by its creators. In order for this system to be of wide use, it must have an abundance of environmental information, at least within certain commercial sectors. This problem is similar to one faced by Wikipedia, RateMyProfessors, and many other community-generated content sites. In order to address it, it is critical that the act of contributing be inherently satisfying. Wikipedia makes people feel like experts, RateMyProfessors lets people vent or commend their teachers, and Urban Dictionary allows people to feel like part of the "in" crowd; GreenScanner could be improved by making the process of contributing environmental information inherently satisfying. One way in which it could do so is to draw on the example of American Red Cross blood drives, which make people feel good because of their contribution to society. Similarly, sites such as GreenScanner should be designed to help users feel that they are making a real difference when they contribute reviews or change their shopping habits after using the system.

Future Work

GreenScanner was an early project in a growing area that seeks to provide consumers with access to environmental information both at home and at the point of purchase. The system was not without its flaws; nevertheless, thoughts on how to address these issues may be constructive for future efforts in this direction. Improvements to Green-Scanner that may be applicable to similar systems have been discussed over the past several years by various members of my research group, in particular Joel Ross and Ellen Eramya, and are discussed below.

A possible way to improve the value of users' contributions to the database would involve building a system to facilitate the making of comparisons between different products, product categories, companies, and other elements. In using the initial version of the site, users seemed reluctant to provide absolute evaluations of various items. Shifting the focus to comparisons could help people feel more capable of contributing meaningful information. From the comparisons that people enter across different aspects of the database (e.g., products or companies), it would become possible to compare many other elements indirectly. For instance, if a certain can of corn is produced by the company Green Giant, and Green Giant is compared favorably to Bird's Eye, then the Green Giant corn is likely to be better than a can of corn by Bird's Eye. While the reliability of these comparisons might not be strong initially, they would provide a basis for other users to refine the database over time.

Another mechanism for bootstrapping the database with information about a wide variety of products would come from analyses of ingredients in each product. There exist various databases online and elsewhere of the ingredients in each product. By enabling people to rate and compare specific ingredients, the site would be able to provide initial valuations of many different products. Analyses of a few commonly used ingredients would permit some information to be supplied for many thousands of different products. "High Fructose Corn Syrup," for example, is found in about 10 percent of the products in one popular online database.

There could also be a reputation management system to allow providers of high-quality content to have a greater impact on the system. Anyone could use GreenScanner to contribute statements or simply click icons to evaluate those statements. In addition, users would be given the opportunity to create an account with the site. All registered contributors would have a reputation score based on the evaluations that other people had given to the environmental impact statements that they had posted. This score would be based on the sum of the review quality scores for all reviews contributed by that user. Also, just as Slashdot allows people to modify whose comments to see based on posters' reputations (Lampe, Johnston, & Resnick, 2007), GreenScanner could allow different individuals' comments to be included or excluded based on their reputation scores. The version of GreenScanner currently online includes a simple reputation management system.

An approach that would dovetail with any of the above efforts would involve using Amazon's Mechanical Turk or a similar system to help fill out the database. Through Mechanical Turk, large numbers of people can be enlisted to assist with small tasks for small amounts of money. Turkers (as the users are called) are willing to perform tasks similar to ranking the environmental impacts of products for only a few U.S. cents and write product reviews for less than a dollar. For several thousand U.S. dollars, an organization could populate an environmental database with information about the most popular products relatively quickly. The quality of these reviews would vary, but it would at least provide a baseline of public opinion from which to build.

Potential Impact

The primary objective of GreenScanner is to help consumers engage in environmentally preferable purchasing. A pivotal moment in a consumer's decision-making process is at the point of purchase, where a choice between two or more products may hinge on small differences among them. By letting people scan the UPCs of products using the camera on their mobile phone and retrieve information from an online database of environmental reviews, GreenScanner helps shoppers figure out what to buy.

In 2004, the U.S. Bureau of Labor Statistics (2006) reported that there were more than 116 million "consumer units" in the United States, each consisting of an average of 2.5 individuals. It may be estimated that the percentage of these consumer units that carry mobile phones while shopping is high, since there were 255 million mobile phone subscribers in the United States in 2007 (Central Intelligence Agency, 2009) or 1.1 subscriptions for each of the 228 million people over 18 years old in the United States (U.S. Census, 2009a). While not all phones have cameras and data service, if 5 percent of all consumer units were to access GreenScanner while making their consumption decisions, this would result in 5.8 million consumer units being influenced by this process.

Each of these consumer units enjoyed an average income before taxes of US$54,453, with US$43,395 in average annual expenditures (U.S. Bureau of Labor Statistics, 2006). Approximately 13.2 percent of each consumer unit's annual expenditure was spent on items that are likely to have UPCs. These items included "Food at home" (7.7 percent), "Alcoholic beverages" (1.1 percent), "Personal care products and services" (1.3 percent), "Reading" (0.3 percent), "Tobacco products" (0.7 percent), and "Apparel and services" (2.1 percent). While some items in each of these groups may not have UPCs, items in several other categories may have UPCs as well. This percentage amounts to US$5,730 of UPC-labeled items per consumer unit per year.

If 5.8 million consumer units consulted GreenScanner in these expenditures, a total of US$33 billion in purchases would be impacted. Altering the decision process even slightly for this volume of purchases could have a profound impact on the total amount of solid waste produced.

For example, if GreenScanner affected 10 percent of a consumer unit's purchasing decisions, and the difference in impact were a 10 percent reduction in waste from the old product to the new product, then GreenScanner would have reduced that consumer unit's total consumer waste by 1 percent. If each consumer unit includes 2.5 people, 5.8 million consumer units use GreenScanner, and each person produces 0.54 kilogram of UPC-labeled waste per day, then GreenScanner would reduce consumer waste by more than 28 million kilograms per year.

Discussion   Many consumers could be willing to change their purchasing habits to take environmental issues into account. However, these individuals often do not have sufficient information to allow environmental concerns to factor into their decisions at the point of purchase. Systems such as GreenScanner can help these consumers engage in environmentally preferable purchasing in some or all of their everyday purchasing decisions. By enabling consumers to have environmental impact information at their fingertips while shopping, GreenScanner and similar efforts contribute to a number of environmental goals, such as the reduction of consumer waste.

Systems such as GreenScanner aid in keeping humans at the center of environmental issues, focusing especially on the impact of consumer decisions. The levels of human consumption in general are problematic, as are the specific choices that people make about what to consume. Situating environmental decision making at the moment of purchase can focus people's attention on the problem at one of its most critical points. The vision for this system is to provide a forum for exchange of environmental information in a format that is reliable and easy to access. By doing so, people around the world may be enabled to make more informed decisions in their everyday lives, and potentially work together to address the environmental issues that surround each of their lives.

During its public deployment, the GreenScanner system gathered a small but stable user community, yet shortcomings in the design and development prevented that community from growing and making the necessary subjective contributions that could allow the site to have a broad environmental impact. Other systems such as GoodGuide (goodguide.com, 2009) and *Consumer Reports'* Greener Choices (greenerChoices.org, 2003) are doing excellent work at gathering databases of environmental information. Mobile device applications are beginning to emerge in this domain as well, such as Seafood Watch (montereybayaquarium.org, 2009). Ultimately, these kinds of systems could reach beyond the limits of only UPC-labeled products to include any systematically identifiable content. For instance, RFID chips, car VIN numbers, and other systems could also be used. While there are numerous questions involved with this expansion, from technical concerns to privacy issues, there may be significant environmental benefits to be gained by collecting environmental information about objects. Hopefully in the coming years a system will be developed that delivers on GreenScanner's original model—a mobile application that provides comprehensive and accurate environmental information about many different products at the point of purchase.

EHCC Perspective

Systems for enabling communities to understand environmental issues and take action together provide good examples of the EHCC perspective. They show how the improvement of our abilities to think and work collectively thereby helps augment our capacity for operating on broader scales of time, space, and complexity.

The GreenScanner project, for example, helps people extend their time horizons in two main ways. First, it gives users of the system a better understanding of the environmental impact of the purchases they make, thus connecting a short-term decision (what to buy) with long-term impacts (how much waste). Second, it encourages contributors to the database to lengthen their time horizon for doing research, helping them know that the benefits of their research will extend beyond their own personal use at this moment, and enabling other people to make more informed decisions in the future. Many other community-based systems provide similar advantages across longer scales of time.

Collective technologies also broaden the spatial aspects of human action. They enable people to perform tasks across geographic distance in a synchronized manner. This coordination can occur in nearby locales, as members of the same community cooperate on joint tasks, or across great distances, as strangers from vastly different locations work together in real time or asynchronously. In addition to helping spatially distributed groups of people act together, enabling people to realize that actions with broad spatial and social extent are even possible is a profound impact of these technologies.

The ability to help people manage social and disciplinary complexity is another key benefit of technologies for collective action. Not just enabling people to work together but enabling *very different* people and various institutions to work together—entities with different goals, ideas, ideals, and abilities—is of critical importance. Coordinating large teams has traditionally been a difficult challenge; collective technologies make it possible for people without a personal fortune or standing army to bring together many others to perform great things.

Humans are social animals; enabling us to work together socially is a central need in an effort to launch a large-scale response to Earth's growing environmental concerns. Green IT can help this effort, helping people find like-minded individuals, share ideas, and work together to accomplish far more than each could alone. With the spread of both desktop and mobile computing around the world, IT is having a significant impact on the ways people collaborate, and is poised to continue transforming human communication and cooperation. Harnessing these capabilities for environmental ends is critical.

9 Ways Forward

If our black and nervous civilization, based on coal, shall be followed by a quieter civilization, based on the utilization of solar energy, that will not be harmful to progress and to human happiness.

—Giacomo Ciamician, 1912

Humans have a long history of solving problems that limit our ability to thrive. We built structures to protect ourselves from the elements. We developed agriculture to guard against vagaries in the availability of wild food sources. We developed codes of laws to organize ourselves. We cured diseases that threatened our communities. Each time some threat has arisen, either externally or internally, humans have risen to meet the challenge. The evidence for our success in this regard is the great number of humans currently alive.

Our ability to respond to these hardships has been magnified by our skill with tools and techniques for taking action, both individually and together. The brick, plow, pen, and microscope are all products of human ingenuity that have allowed us to shape the contexts in which we live—our "environment"—in one way or another. Working together, we have developed ever more complex innovations, building on what has gone before.

Our very success, however, has created the next great challenge. As was discussed in chapter 2, a great deal of evidence now suggests that humanity is on an unsustainable trajectory—that we are using too many resources and producing too much waste for the global civilization to continue indefinitely without profound changes. How we respond to this challenge will be of great importance over the coming decades, centuries, and millennia.

This challenge is unique in its breadth. It is the result of the collective actions people have taken over thousands of years and around the world. It is a side effect of actions that were taken for millions of other reasons—good reasons, certainly, at the time they were taken—such as feeding our families, helping those less fortunate than ourselves achieve a better quality of life, and creating beauty. The collective impacts of these actions, however, are now becoming deeply problematic.

In order for humanity to respond to these challenges, we must develop ways to think about the effects of our actions across long time scales, bridge across great distances, and integrate our capabilities in vastly different fields. Unassisted, humans are not skilled at dealing with problems this great. Fortunately, we have the capability to develop technologies that can help us understand and act at scales beyond our own. This is the same capability that has allowed humanity to transform Earth's ecosystems to date.

IT in particular can help us extend our reach, enabling us to communicate with and learn from millions of other people, see millions of miles, discover what happened millions of years ago, and predict the distant future. IT can help people and organizations develop an understanding of the actions that must be taken here and now in order to have broad effects. It can integrate personal and institutional efforts, and connect them both to environmental ramifications. By helping humans collect and process information, communicate with each other, and control systems with precision, IT is well suited for coordinating a coherent set of responses to global environmental concerns. By broadening our horizons, IT can help us understand the full impact of our past actions and full potential of our future choices.

This book has presented a number of ideas that can facilitate a scientifically grounded vision of a future in which humans and other species continue to exist indefinitely. A major challenge in this process involves understanding how new approaches to sustainability integrate with the existing realities of the world. This chapter considers this integration process as a way of tying together these ideas. By fostering understanding of the ways in which humanity can move from its current state of being to a more sustainable future, and how IT can facilitate this transition, this book can contribute to a long-term resolution of the environmental challenges with which we all are currently faced.

Enacting Change

If we have a vision of sustainability, how do we cause the world to enact this vision? Reality is full of SUVs, Styrofoam, and social structures that will not simply vanish at the snap of one person's fingers. Rather, we need to understand the stages between the current reality and a sustainable future, and then think through a path between them. There are two main ways in which changes become integrated: continuous gradual change and discrete quantum transformations.

Continuous Gradual Change

At the Greening of the Internet Economy conference, panelist Will Recker told of an intelligent transportation system prototype in which autonomous computer-controlled cars were able to travel sixty miles per hour, two to three feet apart. This system was not able to achieve broad usage, though, because of the lack of interim steps from a fully human-driven transportation system to a fully autonomous system; a partial solution was nonviable. As Recker (2009) summed it up, "Any new system [has] to have some way of getting from here to there."

Envisioning an internally consistent world order that is sustainable in the long term is challenging enough in itself. However, this challenge becomes even more formidable when presented with the fact that the world needs to change from its current reality to this proposed future in some series of steps. While a powerful government with visionary scientists, social engineers, land-use planners, and other experts may have the capacity to enact sweeping changes over a relatively short period of time, there are many unavoidable issues that must be dealt with in order for global transformation to occur. For example, even if people were to abandon many of the roads currently being used around the world, those roads would still exist, and would continue to have an impact on ecosystems for decades or longer. People can take action to transform "existing situations into preferred ones" (Simon, 1996), but that action requires prior research and understanding to be effective.

There is a story that my grandfather, who was the chief accountant for Scott Paper Company, used to like to tell. In this story (which I found out later is derived from an Indian folktale involving a girl, a raja, and some grains of rice), a young boy does a favor for a king. The king, in return, asks him what reward he would like. The boy says he would like just one penny today and then two pennies tomorrow, and for his reward to keep doubling each day, until a month had elapsed. The king thought nothing of such a small request and granted it happily, considering the boy to be naive in asking for so little. By the end of the month, however, the boy had all of the money that had been in the king's coffers.

This story is relevant to the environmental topic at hand in that, as an initial step toward global transformation, humans could begin moving toward sustainability through a commitment for continuous and cumulative smaller changes. Restructuring of the world's ideology takes time; as that restructuring gradually comes about, people might initially be encouraged to make a series of modifications to their lifestyles, with the goal that these incremental changes, like the boy's coins, will add up over time. Simultaneously, through institutional and infrastructural change, we might support these smaller changes on a larger scale.

In mathematics, a monotonic function is one in which the direction of change is constant. For example, one's age is a monotonic function (since it is always increasing), whereas one's weight is not (since it can go up or down). The goal of this section is to strive for monotonic change in people's environmental impact. While it may not be possible for a person or organization to become sustainable instantaneously, those entities can nevertheless seek to become better a bit at a time, and never to backslide. Continuously improving one's environmental efforts is valuable in a number of ways. First, the behavioral change has a direct impact on the state of the world. Second, it establishes a lifestyle choice that leads one to watch for additional opportunities to live sustainably. Third, it helps provide other people with possibilities for their own lifestyle modifications.

Technology will be a critical component in the effort to improve the sustainability of human civilization. IT, specifically, provides powerful tools to help people adopt environmental perspectives and lifestyles. IT systems can encourage people to engage in monotonic changes with regard to environmental issues. Many of the projects and ideas presented here have dealt with this sort of incremental improvement. From Trackulous, which gives people tools for keeping track of their lives, to the Wattson device (diykyoto.com, 2009), which monitors home energy usage, to the various carbon offset programs, which allow people and organizations to compensate for unsustainable choices, IT systems are contributing in many different ways to the possibility of continuous gradual change.

Fortunately, many of these new approaches to sustainability use technologies that already exist. There are a number of benefits to basing an environmental intervention on an existing platform. First, it enables the intervention to take advantage of existing infrastructure, rather than needing to pioneer both the intervention and the platform simultaneously. Second, because of the breadth of existing market penetration (e.g., three billion mobile phones), a project might be able to have much broader reach due to the established underlying platform. Third, people may feel more comfortable with an existing piece of technology, and therefore may be more willing to consider environmental activities using it. Finally, since infrastructure takes resources to deploy, the intervention can be undertaken with less marginal environmental impact than an effort that needs to build out its own infrastructure as well.

In some cases, however, the infrastructure may not exist to permit a particular environmental intervention. If the need is great enough in these cases and the awareness of it is sufficiently widespread, there may be a social impetus to set up entirely new infrastructures. For example, there is currently an effort afoot in the United States to bring about an infrastructure of charging stations for electric cars as an alternative to gas stations. Nevertheless, the effort required to enact change at this scale is nontrivial.

The broad goal of many Green IT approaches is to encourage us to look for opportunities for incremental environmental improvement in all aspects of our lives. If we are already using technological systems, then we should be able to use those systems to enact sustainable change as well. Incremental improvement across many people and activities can collectively have a transformative effect.

Two main factors contribute to a continuous gradual movement toward environmental sustainability: personal change and broad social adoption. At an individual level, each person may engage in gradually different behavior; so too may change spread at broader social and institutional levels, as more and more people and organizations take on a behavioral pattern. These two factors combine to enable societies as a whole to have a great impact through continuous gradual change.

Social forms of Green IT are critical to global sustainability efforts. The value in these efforts does not lie solely or even primarily in the difference made by individual lifestyle changes. A large part of the value of social Green IT systems rests in their ability to change individuals' minds and encourage them to identify as environmentally aware individuals. When these people assume positions of power within existing organizations, take up the mantle to oppose social problems, raise their children, or enact the thousands of other possible behavioral patterns that make up a human life, they will imbue these actions with environmental awareness. This awareness will be the stuff of which social transformation is made.

People need both emotional engagement with sustainable ideals and accurate information on which to act. To create emotional engagement, people would benefit from being situated at the center of sustainable issues, but not at the center of a huge guilt trip. We might see environmentally inspired change as an opportunity to improve our lives, rather than as a burden thrust on us or a stripping away of resources. As the myriad projects described in chapters 6–8 sought to make clear, IT can help bring these forms of emotional engagement to pass.

Providing accurate information about how best to change and reliable documentation about change that has already occurred is critical as well. IT can help establish metrics for determining the success of various efforts, measuring observables, and disseminating the results. It can help people know that the efforts being undertaken are actually working, and that their contributions are having real impacts. Both emotional engagement and accurate information will assist people in gradually deepening their involvement in environmental concerns.

This section has discussed the need for continuous gradual change across many different people's lifestyles to achieve environmental sustainability. Still, in some circumstances successive refinement can only take one so far. New research discoveries provide opportunities for major behavioral or social alterations that can achieve profound transformations. There may be steps along the path to sustainability where successive refinement is not viable and some quantum leap is necessary.

Quantum Change

Gradual changes to the world's current way of functioning might be the easiest way to encourage broad-scale participation in environmental efforts. However, there is precedent for major social transformations happening in a relatively short period of time. Revolutionary social movements have shaped the modern era, with new governments taking over and dramatically altering the fabric of society in those countries. The collective shift in behavior and perspective that occurs in countries as a result of cataclysmic events such as war or natural disasters demonstrates the willingness of groups to work together to address a common cause. Technological innovations such as electricity and computing have also ushered in rapid changes on a large scale around the globe. For example, within the last fifteen years, the Internet has caused significant social shifts around the world, transforming commerce, communications, entertainment, politics, and many other fields. These transformations, facilitated by political, social, technological, and other causes, play a major role in the way civilizations operate.

It is possible to think about the relationship between gradual change and quantum transformations as being like the different the gears of a car. When a car shifts into a higher gear, it may suffer reduced efficiency or speed momentarily, but is able to go faster in the long run. So too with the environmental movement: many gradual changes collectively move humanity toward sustainability, but occasionally we may need to undergo more dramatic alterations in order to be able to make gradual progress once again.

There is a significant concern that gradual change may not be sufficient to achieve global sustainability. As the *Smart 2020* report asserts, "It is becoming clear that incremental change is not going to be enough to tackle climate change to the degree and at the speed required to keep carbon at 'safe' levels in the atmosphere" (Climate Group, 2008, p. 51). IT can be of use here as well, helping us recognize discontinuities between business as usual and sustainable lifestyles.

In his book *The Structure of Scientific Revolutions*, Thomas S. Kuhn (1996) discusses the processes through which science undergoes paradigm shifts. While everyday science helps to flesh out the details of new paradigms, Kuhn proposes, it does not lead to paradigm shifts in science; rather, shifts are caused by crises. As Kuhn notes, "The significance of crises is the indication they provide that an occasion for retooling has arrived" (p. 76). It is likely that one or more paradigm shifts will occur as the scientific community discovers better ways to think about and address the world's environmental concerns. These shifts are likely to trigger vastly new kinds of social responses to the environmental crisis.

Even within existing scientific paradigms, the tipping point phenomenon could bring about dramatic social transformations in a short period of time. The core premise here is that once that point is passed, the ensuing change is self-reinforcing and thus escalates. Peak oil is an example of a tipping point that impacts the world's environmental issues. Because so much of the world's energy infrastructure is predicated on oil and other petroleum products, the point at which oil production peaks is seen as pivotal in terms of the humanity's energy supply. The decline of energy availability after peak oil could lead to not just high gasoline prices but also problems in many different facets of society. According to CNN, "To feed an average family of four in the developed world uses up the equivalent of 930 gallons of gasoline a year—just shy of the 1,070 gallons that same family would use up each year to power their cars" (Oliver, 2008). Because of the far-reaching impacts of decreased oil production, there will likely be revolutions in not just transportation but also food production, manufacturing, and many other fields.

The alternative to revolutionary changes, in some cases, may be catastrophe. Despite this looming threat, the ability of human cultures to respond vigorously to transformative events in the past offers hope that we will develop profoundly new ways to work together, supported by and utilizing IT, that help us respond to this new set of concerns.

Balancing Research and Action

One of the challenges faced by anyone desiring to respond to environmental issues is in knowing what to do. We have imperfect information about our world, and that limits our ability to make perfect plans to respond to it. Nevertheless, the lack of exact accuracy in our understanding does not relieve the pressing need for a response. Action in the face of uncertainty is difficult and sometimes scary, but it is necessary when the uncertainty involves a non-negligible risk of worldwide cataclysm.

While scientists are able to make increasingly accurate models, it is not possible to have perfect information on climate change and other environmental issues. Despite this limitation it is still important that people take action to address these issues, given the high likelihood that the global environmental situation is becoming increasingly problematic, and that humans are in large part the cause. Research is critical to understanding the problems at hand, but it is ultimately only as effective as the people who act on that research to produce change.

In order to have the greatest impact on present-day environmental issues, we must strike a balance between research to understand the problems and acting to effect their solutions. If we spend all of the available effort on research, without any attention to acting on the results, no actual change in environmental impact will occur. In fact, a call for deeper understanding is sometimes used as a technique for preventing action. On the other hand, action alone, based on the current imperfect information that we have about these issues, may result in misdirected effort, solving trivial problems, missing important and achievable goals, and potentially doing more harm than good. Engaging simultaneously in high-quality science and vigorous courses of action will have the greatest effect.

I once thought that Socrates, Plato, and Aristotle were philosophers who were famous purely on the strength of their own ideas. However, my understanding of the situation changed dramatically when I discovered that Aristotle had been the childhood tutor of Alexander the Great. While perhaps Socrates, Plato, and Aristotle would have been remembered even had this not been the case, I find it much more likely that they joined the ranks of the foundational thinkers of Western civilization because Alexander conquered Greece, Egypt, most of the Middle East, and a large chunk of Asia, and spread their ideas in his wake. In this spirit, research is rendered important when people choose to act on it. Until we do so, research may have a great deal of potential, but it will not change the world on its own. This book has sought to propose tools that can enable people to work together to turn the ideas of researchers into positive environmental change, thereby becoming, collectively, a kind of distributed Alexander the Great for the environmental movement.

Enacting environmental change will not be easy. Change in itself is often difficult, and some people and institutions oppose environmental action. Entities that resist action around environmental issues do so for a number of reasons. They may stand to lose a great deal of money from the changes that would be necessary, as the oil industry might. They may prefer to devote their energies to other issues that they find more pressing, such as politicians supporting economic growth. In some cases, they may even be looking forward to the end of the world, as in the case of the religious group that created the Rapture Index (raptureready.com, 2008).

Some groups that oppose environmental action may not do so openly. Given the prevailing culture in the United States and many other countries, it would be socially and politically problematic to oppose environmental issues explicitly. Therefore, opponents of environmental action often claim that there is not enough information to act effectively, and that we need more research before we can act. U.S. president George W. Bush (2002), for example, in a speech on climate change, said, "When we make decisions, we want to make sure we do so on sound science; not what sounds good, but what is real." President Bush was not renowned as a champion of environmental concerns during his presidency; his calls for "sound science" have been seen by some (including Kofi Annan, then secretary general of the United Nations [Kirby, 2001]) as a way of delaying action in the present.

More recently, however, there has been greater support for environmental action at the highest levels of the U.S. government. Shortly before taking office, then President-Elect Barack Obama made the following statement:

Now is the time to confront this challenge once and for all. . . . Delay is no longer an option. Denial is no longer an acceptable response. The stakes are too high. The consequences, too serious. Stopping climate change won't be easy. It won't happen overnight. But I promise you this: When I am president, any governor who's willing to promote clean energy will have a partner in the White House. Any company that's willing to invest in clean energy will have an ally in Washington. And any nation that's willing to join the cause of combating climate change will have an ally in the United States of America. (Youngman, 2008)

Ultimately, it is necessary to undertake simultaneous efforts in research and action, accompanied by a willingness to accept a degree of error, when the decisions causing that error were based on the best available information at the time. Tightening the feedback loop between researchers and action takers can help to reduce the margin of error by getting the most accurate information to the people taking action as quickly as possible. IT can provide tools and techniques for making this feedback loop more rapid, thereby reducing the time lag and mitigating the effects of any potential errors.

The EHCC approach can support this balance between research and action. By considering the scales of time, space, and complexity of the environmental problems and relevant human responses, designers may be able to understand how IT systems can narrow the gap between the two. Evaluators of systems can use this conceptual framework to assess how effectively these systems perform at bridging the two scales. Modelers and policymakers may use this approach to consider where problem spots are likely to lie in an effort to achieve sustainable targets. Across many different efforts, this approach can also be used to establish connections and develop taxonomies. The Green IT systems developed using an EHCC approach, and the ways in which existing systems are adapted for environmental ends, can collectively help humanity live more sustainably.

Long-term Impacts

A professor at MIT named Hiroshi Ishii, while giving a guest lecture to a class of first-year PhD students, once provided the following exercise. He first drew a horizontal time line on the whiteboard. Then he drew a short vertical line at the left end of it and wrote "2007" above it. Next, he drew a vertical line to the left of center of it and labeled it "2050." Above it, he drew a little tombstone with "RIP: me" on it. Next, he drew a line at 2100, again with a tombstone, and wrote "RIP: all of you" on it. Then he drew a line at 2200 and asked, "What will the impact of your research be here?"

While the task of planning a research agenda for its impact two hundred years in the future is a bit daunting, it nonetheless opens up a broader horizon than most people usually consider. Corporate research and development tend to deal with issues one to five years in the future; academic research typically deals with events two to ten years away. Thinking about impact twenty times further into the future requires a different set of approaches. The Long Now Foundation has made efforts in this direction, beginning to think about problems on a time scale of ten thousand years—for example, by adding a 0 to the year (02009) to avoid the "deca-millennium bug" that will occur in a bit less than eight thousand years.

This book has chosen many of its case studies around shorter-term phenomena than those that are two hundred or ten thousand years in the future, in part to help create a connection with broader audiences. However, these instances can begin the process of extending our horizons. By looking at the ways in which IT can broaden the scales of time, space, and complexity that we consider when making decisions, the ideas presented here may be of some use in stretching our attention beyond our proximate concerns.

IT has been around for a long time—from number systems to written text to maps to the printing press to computers. Pressing social and cultural issues have been around just as long as IT. From diseases to warfare to the need to stay warm and find food, humanity has been faced with many formidable challenges. Environmental issues are perhaps the most critical of the challenges that humans face, but there are many others. Just as IT has supported efforts to deal with past challenges, it will be a key component of our response to environmental issues as well.

Other Global Challenges

Environmental sustainability has been the reason for discussing the broadening of horizons through IT in this book. Looking beyond this set of problems, though, there may be other applications for the perspectives presented here. The United Nations offers eight "Millennium Development Goals": ending poverty and hunger, ensuring universal education, promoting gender equality, reducing child mortality, improving maternal health, combating HIV/AIDS, achieving environmental sustainability, and developing a global partnership (United Nations, 2009). As with sustainability, the seven other goals have a certain commonality: they are all beyond the scope of any one individual. Still, they are all critical to a just future.

Many of the lessons learned in the close analysis of IT for environmental sustainability could be applied effectively in these other problem domains. Looking for gaps between the time, space, and complexity horizons of individual humans and the scale of the problems, and seeking ways for IT to bridge those gaps, could help pave the way toward novel approaches to these overwhelming issues. Understanding the disconnects of scale that are causing systemic problems and helping to remediate those concerns could be among the many contributions of IT to addressing global issues. While IT is not a panacea, it is an enabler of diverse human undertakings. If humans have the power to fix the vast problems facing us, it is likely that we can do so more quickly through the use of IT.

A Cautiously Optimistic Perspective

In light of the many complex environmental challenges confronting humanity today, it is clear that the thousands of projects currently under way around the world and the several hundred pages of this book only begin to scratch the surface of the problem. Nevertheless, I remain cautiously optimistic about humanity's ability to respond to these challenges without needing to survive an apocalypse. Cautious, because the challenges are great and the stakes are high. Optimistic, because what humans lack in foresight, we make up for in enthusiasm and compassion. While it would be far more efficient to enact small changes over a longer period of time, we are instead setting ourselves up to need a much more urgent response to more dramatic problems. The transition would probably be easier if everyone were able to take a longer time horizon into account while making decisions. The value of a long time horizon, however, is predicated on people having access to basic necessities that allow the luxury of looking to the future; the education to understand what is coming and have ideas about how to address it; and the social, corporate, and governmental support to take action in response. Nevertheless, even with the relatively short-term view shared by most people (and I do not exempt myself; if my computer is not working correctly or I cannot find a parking space, environmental issues rapidly move to the back of my mind), at some point the issues will become sufficiently pressing that we will be forced to confront them.

Every effort I can think of to contribute to global environmental sustainability involves some degree of environmental cost as well. Moving to a more sustainable power infrastructure involves disposing of the materials from the previous system. Bicycling to work instead of driving means one needs to eat more food, which has a nonzero carbon footprint. Even the time spent thinking about what environmental actions to take involves costs, if only just the opportunity cost of not taking some other action during that time. The goal in any of these cases is simply to have the benefits outweigh the drawbacks.

Consider, too, this book. It is printed on dead trees, albeit recycled dead trees (except for those of you who may be reading it online or as an e-book). Nevertheless, it seeks to have a positive net environmental impact. The archival value of a paper-based book, and the quality control, credibility, and distribution apparatus of a major university press can all help to broaden the impact of the ideas. Although it involves some environmental costs, hopefully the benefits are greater.

I wrote this book to help set the stage for the social transformations that I see coming in the near future, by addressing the role that IT can play in the change of humanity's style of living from unsustainable to sustainable. Hopefully thereafter, IT can support us as we move beyond sustainability to the restoration of ecosystems and new kinds of abundance. While IT and humanity's relationship with it are not without their shortcomings, such as e-waste, power usage, and irrational exuberance about how much it can help, there is a great deal of potential in using computational tools to manipulate and share information. This book has sought to provide suggestions of how IT can facilitate opportunities to change individually and connect with others to live together more sustainably. Excitingly, the payoff will not just be in the continuity of life on Earth but also in closer connections within our communities and with the world around us.

Humanity faces vast environmental challenges in the coming years; yet to suggest that they are unconquerable would be to place too little faith in the ability of people, both as individuals and as groups, to understand the world and respond to it effectively. Great change is coming; I hope and believe that people can take the initiative to enact change on a global scale, rather than becoming the victims of global change.

Notes

1. Portions of this chapter are reprinted from Tomlinson (2009).

2. This chapter includes text and ideas derived from Tomlinson (2008b).

3. This project was developed by a large team of researchers led by the author and University of California, Irvine ecology and evolutionary biology professor Lynn Carpenter. The core team included Eric Baumer, Grace Chiang, Sara Goetz, Bryant Davey, Stefan Marinov, Calen May-Tobin, Uel McMahan, Robert Moodey, Ed Niecikowski, Riley Pratt, Erika Ramos, Martin Schmidt, Man Lok Yau, Craig Yoho, and Kristin Young. Portions of this chapter are based on research previously presented by the author and his colleagues (Tomlinson, Baumer, Yau, Carpenter, & Black 2008; Tomlinson et al., 2009).

4. This project was developed by the author in 2007. The ideas were improved by discussions with Rebecca Black, Andrew Zaldivar, Rob Olson, Jeff Greenberg, and Lilly Irani.

5. This project was initially developed by the author, and was improved through work by Joel Ross and Ellen Eramya. Portions reprinted, with permission, from Tomlinson 2008a. "Prototyping a Community-Generated, Mobile Device-Enabled Database of Environmental Impact Reviews of Consumer Products," Proceedings of the 41st Annual Hawaii International Conference on System Sciences (HICSS 2008), pp. 154ff. © 2008 IEEE.

References

adventureecology.com. (2009). Adventure ecology. Retrieved February 17, 2009, from <http://missioncontrol.adventureecology.com/>.

Akmam, W. (2002). Women's education and fertility rates in developing countries, with special reference to Bangladesh. *Eubios Journal of Asian and International Bioethics, 12*, 138–143.

archives.gov. (2009). NARA | The National Archives experience. Retrieved March 7, 2009, from <http://www.archives.gov/exhibits/charters/constitution_transcript.html>.

Aveda. (2009). Renewable energy. Retrieved February 27, 2009, from <http://aveda.aveda.com/aboutaveda/renewable_energy.asp>.

Bannon, L. (1992). From human factors to human actors: The role of psychology and human-computer interaction studies in system design. In J. Greenbaum & M. Kyng (Eds.), *Design at work: Cooperative design of computer systems* (pp. 25–44). Hillsdale, NJ: Lawrence Erlbaum Associates. Retrieved May 4, 2008, from <http://portal.acm.org/citation.cfm?id=125458>.

banterminator.org. (2009). Ban Terminator. Retrieved March 1, 2009, from <http://www.banterminator.org/>.

Barab, S., & Dede, C. (2007). Games and immersive participatory simulations for science education: An emerging type of curricula. *Journal of Science Education and Technology, 16*(1), 1–3.

Basel Action Network. (2008). *The digital dump: Exporting high-tech re-use and abuse to Africa.* Retrieved February 16, 2009, from <http://www.ban.org/films/TheDigitalDumpTrailer.html>.

bbc.co.uk. (2005, January 27). Tsunami aid: Who's giving what. BBC. Retrieved May 5, 2008, from <http://news.bbc.co.uk/2/hi/asia-pacific/4145259.stm>.

bbc.co.uk. (2008a). Science and nature: Games. BBC. Retrieved May 5, 2008, from <http://www.bbc.co.uk/sn/games/>.

bbc.co.uk. (2008b, November 10). Plan for new Maldives homeland. BBC. Retrieved November 12, 2008, from <http://news.bbc.co.uk/1/hi/world/south_asia/7719501.stm>.

Becker, B. (2008, April 27). Bicycle-sharing program to be first of kind in U.S. *New York Times.* Retrieved February16, 2009, from <http://www.nytimes.com/2008/04/27/us/27bikes.html?r=1&th&emc=th&oref= slogin>.

behindthegreen.org. (2009). Behind the green: About DESC. Retrieved February 17, 2009, from <http://www.behindthegreen.org/about/desc/>.

Bembenutty, H., & Chen, P. P. (2005). Self-efficacy and delay of gratification. *Academic Exchange Quarterly.* Retrieved February 17, 2009, from <http://findarticles.com/p/articles/mi_hb3325/is_4_9/ai_n29236299>.

Bergman, B. (2008). Green speed dating: Finding carbon-neutral love. National Public Radio. Retrieved February 17, 2009, from <http://www.npr.org/templates/story/story.php?storyId=92537810>.

Berry, M., Hazen, B., MacIntyre, R., & Flamm, R. (1996). Lucas: A system for modeling land-use change. *Computational Science and Engineering, 3*(1), 24–35.

Black, R. (2008). *Adolescents and online fan fiction.* New York: Peter Lang Publishing.

Blevis, E. (2007). Sustainable interaction design: Invention & disposal, renewal & reuse. In *Proceedings of the SIGCHI Conference on Human Factors in Computing Systems* (pp. 503–512). San Jose, CA: ACM.

Blevis, E., & Huang, E. (forthcoming). *Do Less with Design: Foundations of Sustainable Interaction Design.* Cambridge, MA: MIT Press.

BottledWater.org. (2006). *Bottled water continues as number 2 in 2006.* Retrieved August 1, 2009, from <http://www.bottledwater.org/public/Stats_2006.doc>.

Bowers, C. (1990). How computers contribute to the ecological crisis. *CPSR Newsletter, 8*(3), 6–8.

Bregman, J. I., & Lenormand, S. (1966). *The pollution paradox.* New York: Spartan Books.

Bruce, M. (2009, January). Greening of the Internet economy—Conference. Retrieved February 15, 2009, from <http://greeninternet.calit2.net/>.

Brush, A., Turner, T., Smith, M., & Gupta, N. (2005). Scanning objects in the wild: Assessing an object triggered information system. In M. Beigl et al. (Eds.), *Proceedings of the Seventh International Conference on Ubiquitous Computing* (pp. 305–322). Berlin: Springer-Verlag.

Bush, G. W. (2002, February 14). President announces clear skies and global climate change initiatives. Retrieved May 5, 2008, from <http://www.whitehouse.gov/news/releases/2002/02/20020214-5.html>.

California State Board of Education. (2004, December 10). Science content standards. Retrieved March 3, 2009, from <http://www.cde.ca.gov/BE/ST/SS/documents/sciencestnd.pdf>.

Calit2. (2004). California wireless traffic report. Retrieved February 16, 2009, from <http://traffic.calit2.net/>.

Calit2 & CPUC. (2009). Greening of the Internet economy—Conference. Retrieved February 15, 2009, from <http://greeninternet.calit2.net/>.

Cappiello, D. (2008, November 27). Hot over the environment. *Moscow News.* Retrieved February 16, 2009, from <http://mnweekly.ru/world/20081127/55358937.html>.

Capra, F. (2002). *The hidden connections: Integrating the biological, cognitive, and social dimensions of life into a science of sustainability*. New York: Doubleday.

Carpenter, F. L., Nichols, J., Pratt, R., & Young, K. (2004). Methods of facilitating reforestation of tropical degraded land with the native timber tree, *Terminalia amazonia*. *Forest Ecology and Management, 202*, 281–291.

Carr, N. (2006). Avatars consume as much electricity as Brazilians. Retrieved February 16, 2009, from <http://www.roughtype.com/archives/2006/12/avatars_consume.php>.

carrotmob.org. (2008). carrotmob.org. Retrieved December 6, 2008, from <http://www.carrotmob.org/>.

Carson, R. L. (1962). *Silent spring*. New York: Houghton Mifflin.

cbs.com. (2008, August 17). *CSI*. Retrieved August 17, 2008, from <http://www.cbs.com/primetime/csi/>.

cdiac.ornl.gov. (2009). Carbon Dioxide Information Analysis Center (CDIAC)—Frequently asked questions. Retrieved February 15, 2009, from <http://cdiac.ornl.gov/pns/faq.html>.

Central Intelligence Agency. (2009). *World fact book*. Retrieved August 1, 2009, from <https://www.cia.gov/cia/publications/factbook/geos/us.html>.

Ciamician, G. (1912). The photochemistry of the future. *Science*, new series, *36*(926), 385–394.

CITRIS. (2009). Climate Navigator. Retrieved August 1, 2009, from <http://climatenavigator.org/>.

Clarke, A. C. (1973). *Profiles of the future*. New York: Harper & Row.

Climate Group, The. (2008). *SMART 2020: Enabling the low carbon economy in the information age*. Retrieved March 3, 2009, from <http://www.theclimategroup.org/assets/resources/publications/Smart2020Report.pdf>.

climatesaverscomputing.org. (2008). Climate Savers Computing. Retrieved February 17, 2009, from <http://www.climatesaverscomputing.org/>.

Coase, R. H. (1937). The nature of the firm. *Economica,* New Series, *4*(16), 386–405.

Cohen, J. E. (1995). *How many people can the earth support*. New York: W. W. Norton and Company.

Commoner, B. (1972). The environmental cost of economic growth. *Chemistry in Britain, 8*(2), 52–56.

Connelly, L. (2008). County issues warning to stay out of water. Retrieved February 16, 2009, from <http://beach.freedomblogging.com/2008/11/05/county-issues-warning-for-surfers-and-swimmers-to-stay-out-of-water/3450/>.

CSIRO. (2008). Rockhampton: JM Rendel Laboratory and Belmont Research Station (Qld) (Profile—location). Retrieved February 16, 2009, from <http://www.csiro.au/places/Rockhampton.html>.

Danis, C. M., Viegas, F. B., Wattenberg, M., & Kriss, J. (2008). Your place or mine? Visualization as a community component. In *Proceeding of the twenty-sixth annual SIGCHI conference on human factors in computing systems* (pp. 275–284). New York: ACM.

Das, J. (2006). GreenScanner. SuperVegan. Retrieved March 10, 2009, from <http://supervegan. com/blog/entry.php?id=145>.

Dasgupta, P. (2006). State of the planet 2006. Retrieved February 15, 2009, from <http://www. earth.columbia.edu/sop2006/transcripts/tr_dasgupta.html>.

Dawkins, R. (1990). *The selfish gene* (2nd ed.). New York: Oxford University Press.

Democratic Republic of Congo. (2009). Coltan. Retrieved February 16, 2009, from <http://www. un.int/drcongo/war/coltan.htm>.

Diamond, J. (2008, January 2). What's your consumption factor? *New York Times.* Retrieved May 4, 2008, from <http://www.nytimes.com/2008/01/02/opinion/02diamond.html?ref=opinion>.

Diouf, J. (2001). Ethical issues in food and agriculture. Retrieved February 15, 2009, from <http:// www.fao.org/docrep/003/X9601E/x9601e01.htm>.

Dittmar, P. (2008). *Flight KK to KL—Oil palm and forest.* Retrieved March 4, 2009, from <http:// www.flickr.com/photos/dittmars/2436407084/>. Made available under a Creative Commons license, from <http://creativecommons.org/licenses/by-sa/2.0/deed.en>.

diykyoto.com. (2009). DIY KYOTO. Retrieved March 2, 2009, from <http://www.diykyoto.com/ uk>.

Donn, J., Mendoza, M., & Pritchard, J. (2008). AP probe finds drugs in drinking water. Retrieved February 19, 2009, from <http://abcnews.go.com/US/wireStory?id=4416882>.

Dourish, P. (2001). Seeking a foundation for context-aware computing. *Human Computer Interaction, 16*(2, 3, & 4), 229–241.

Dourish, P. (2008). Points of persuasion: Strategic essentialism and environmental sustainability. In M. Foth., C. Satchell, E. Paulos, T. Igoe, & C. Ratti (Eds.). *Pervasive 2008 Workshop on Pervasive Persuasive Technology and Environmental Sustainability.* Sydney, Australia. Retrieved August 1, 2009, from: <http://www.pervasive2008.org/Papers/Workshop/w2-05.pdf>.

Dunbar, R. (1993). Coevolution of neocortical size, group size, and language in humans. *Behavioral and Brain Sciences, 16*(4), 681–735.

Duvall, M. (2008). A green IT infrastructure case study: Monsanto. Retrieved February 16, 2009, from <http://www.eweek.com/c/a/Green-IT/A-Green-IT-Infrastructure-Case-Study-Monsanto/>.

Dweck, C. (2006). *Mindset: The new psychology of success.* New York: Random House.

Earthday.net. (2008). Ecological footprint quiz. Retrieved May 4, 2008, from <http://ww2.earth day.net/footprint/>.

ecorio.org. (2009). This is ecorio. Retrieved February 17, 2009, from <http://www.ecorio.org/>.

Eldridge, N., & Gould, S. J. (1972). Punctuated equilibria: An alternative to phyletic gradualism. In T. J. M. Schopf (ed.), *Models in paleobiology* (pp 82–115). San Francisco: Freeman, Cooper & Co.

Elliott, H., & O'Connor, G. (2007). Phosphorus management for sustainable biosolids recycling in the United States. *Soil Biology & Biochemistry, 39*(6), 1318–1327.

epinions.com. (2007, June 14). Epinions. Retrieved March 3, 2009, from <http://www.epinions .com/>.

Erlanger, S. (2008, August 17). France reaffirms its faith in future of nuclear power. *International Herald Tribune*. Retrieved February 15, 2009, from <http://www.iht.com/articles/2008/08/17/ europe/17francenuke.php>.

Falk, J. H., Dierking, L. D., & Foutz, S. (2007). *In principle, in practice: Museums and learning institutions*. Lanham, MD: Altamira Press.

Fitzenberger, J. (2006). New online database lets consumers share product information at point of purchase. Retrieved August 1, 2009, from <http://today.uci.edu/news/release_detail.asp?key=1459>.

Food and Fertilizer Technology Center. (2004). Harnessing the potential of IT toward sustainable agriculture in Asia. Retrieved February 16, 2009, from <http://www.agnet.org/library/nc/145a/>.

Foreman, D. (1991). *Confessions of an eco-warrior*. New York: Crown Trade Paperbacks.
Foucher, S. (2009). Peak oil update. *Oil Drum*. Personal communication. Made available under a Creative Commons license, from <http://creativecommons.org/licenses/by-sa/3.0/us/>.

Friedman, B., & Freier, N. G. (2005). Value sensitive design. In K. E. Fisher, S. Erdelez, & E. F. McKechnie (Eds.), *Theories of information behavior: A researcher's guide* (pp. 368–372). Medford, NJ: Information Today.

Gaffney, D. (2007). This guy can get 59 MPG in a plain old Accord. Beat that, punk. *Mother Jones*. Retrieved February 17, 2009, from <http://www.motherjones.com/politics/2007/01/guy-can-get-59-mpg-plain-old-accord-beat-punk>.

Gardyn, R. (2002, September 1). Being green. Retrieved February 17, 2009, from <http://findarti cles.com/p/articles/mi_m4021/is_2002_Sept_1/ai_90957499>.

globe.gov. (2009). GLOBE: Home. Retrieved February 17, 2009, from <http://globe.gov/r>.

goodguide.com. (2009). Ratings of natural products, green products, and healthy products. Good-Guide. Retrieved March 2, 2009, from <http://www.goodguide.com/>.

Google.org. (2009). Plug into a greener grid: RE<C and rechargeIT initiatives. Retrieved February 16, 2009, from <http://www.google.org/rec.html>.

Gottlieb, R. S. (1996). *This sacred earth*. New York: Routledge.

Green, H. (2006). GreenScanner, UPC database for the environment. Clean technology, EcoGeek. Retrieved March 10, 2009, from <http://www.ecogeek.org/content/view/50/27/>.

GreenerChoices.org. (2003). GreenerChoices.org. Retrieved March 2, 2009, from <http://www.greenerchoices.org/>.

Greenpeace. (2007). Habbo and Greenpeace survey reveals teens more concerned about greenhouse gases than drugs, violence, or war. Greenpeace International. Retrieved May 4, 2008, from <http://www.greenpeace.org/international/press/releases/habbo-and-greenpeace-survey-re>.

Griscom, A. (2002, December 19). Activists are split on a proposed wind project off Cape Cod. Grist. Retrieved February 15, 2009, from <http://www.grist.org/news/powers/2002/12/19/griscom-windmill/>.

Griskevicius, V., Tybur, J., Sundie, J., Cialdini, R., Miller, G., & Kenrick, D. (2007). Blatant benevolence and conspicuous consumption: When romantic motives elicit strategic costly signals. *Journal of Personality and Social Psychology, 93*(1), 85–102.

Grossman, E. (2006). *High tech trash: Digital devices, hidden toxics, and human health.* Washington, DC: Island Press.

Grunwald, M. (2002, February 23). Monsanto held liable for PCB dumping. WashingtonPost.com. Retrieved March 1, 2009, from <http://www.washingtonpost.com/ac2/wp-dyn/A54914-2002Feb22?language=printer>.

Guardian News & Media. (2008). Flights reach record levels despite warnings over climate change. Retrieved February 16, 2009, from <http://www.buzzle.com/articles/136859.html>.

Guggenheim, D. (Director). (2006). *An inconvenient truth* [motion picture]. United States: Paramount.

Hardin, G. (1968). The tragedy of the commons. *Science, 162*(3859), 1243–1248.

Hartmann, T. (2004). *The last hours of ancient sunlight: Revised and updated: The fate of the world and what we can do before it's too late.* New York: Three Rivers Press.

Hawken, P., Lovins, A., & Lovins, L. H. (2000). *Natural capitalism: Creating the next industrial revolution.* London: Little, Brown. Retrieved August 1, 2009, from <http://www.natcap.org/images/other/NCchapter1.pdf>.

Hightower, J., Smith, T., Sonnenfeld, D. A., & Pellow, D. N. (2006). *Challenging the chip: Labor rights and environmental justice in the global electronics industry.* Philadelphia: Temple University Press.

Holdren, J. (2008). The science and physical implications of climate change. Retrieved February 15, 2009, from <http://greenmonk.net/john-holdren-on-global-climatic-disruption/>.

Hollis, J., & Weight Loss Maintenance Trial Research Group. (2008). Weight loss during the intensive intervention phase of the weight-loss maintenance trial. *American Journal of Preventive Medicine, 35*(2), 118–126.

Hölzle, U. (2009). Official Google blog: Powering a Google search. Retrieved February 16, 2009, from <http://googleblog.blogspot.com/2009/01/powering-google-search.html>.

Honicky, R., Brewer, E. A., Paulos, E., & White, R. (2008). N-smarts: Networked suite of mobile atmospheric real-time sensors. In *Proceedings of the second ACM SIGCOMM workshop on networked systems for developing regions* (pp. 25–30). Seattle: ACM.

Hopfenberg, R. (2003). Human carrying capacity is determined by food availability. *Population and Environment, 25*(2), 109–117.

Hopper, A. (2008, March 17). Computing for the future of the planet. Lecture at the Royal Society. Retrieved February 15, 2009, from <http://www.cl.cam.ac.uk/research/dtg/~ah12/CFP%20 RS%20slides.pdf>.

Hopper, A., & Rice, A. (2008). Computing for the future of the planet. *Philosophical Transactions of the Royal Society A: Mathematical, Physical, and Engineering Sciences, 366*(1881), 3685–3697.

Hsieh, D. (2009). Greening of the Internet economy—Conference: ICT, efficiency, and the consumer. Retrieved February 15, 2009, from <http://greeninternet.calit2.net/>.

Huang, E. M., & Truong, K. N. (2008). Breaking the disposable technology paradigm: Opportunities for sustainable interaction design for mobile phones. In *Proceeding of the twenty-sixth annual SIGCHI conference on human factors in computing systems* (pp. 323–332). Florence: ACM.

International Energy Agency. (2007). *Tracking industrial energy efficiency and CO_2 emissions.* Retrieved February 15, 2009, from <http://www.iea.org/textbase/nppdf/free/2007/tracking_emissions.pdf>.

International Energy Agency. (2008). *Worldwide trends in energy use and efficiency: Key insights from IEA indicator analysis.* Retrieved March 3, 2009, from <http://www.iea.org/Textbase/Papers/2008/ Indicators_2008.pdf>.

IPCC. (2007a). *Climate change 2007: Synthesis report. Contribution of working groups I, II, and III to the fourth assessment report of the Intergovernmental Panel on Climate Change.* Geneva, Switzerland. Retrieved March 3, 2009, from <http://www.ipcc.ch/pdf/assessment-report/ar4/syr/ar4_syr.pdf>.

IPCC. (2007b). Summary for policymakers. In S. Solomon, et al. (Eds.), *Climate change 2007: The physical science basis. Contribution of working group I to the fourth assessment report of the Intergovernmental Panel on Climate Change.* Retrieved August 1, 2009, from <http://ipcc-wg1.ucar.edu/wg1/ Report/AR4WG1_Print_SPM.pdf>.

Ishii, H., Kobayashi, M., & Grudin, J. (1992). *Integration of inter-personal space and shared workspace: ClearBoard design and experiments.* Proceedings of the 1992 ACM conference on Computer-supported cooperative work. (pp. 33-42). Toronto: ACM.

ISO. (2006). *ISO 14064–1:2006—Greenhouse gases—Part 1: Specification with guidance at the organization level for quantification and reporting of greenhouse gas emissions and removals.* Retrieved February 16, 2009, from <http://www.iso.org/iso/iso_catalogue/catalogue_tc/catalogue_detail.htm?cs number=38381>.

IT-E3. (2004). IT-E3 tools home. Retrieved May 5, 2008, from <http://ties.ucsd.edu/ITEST/index. html>.

Ito, M., Baumer, S., Bittantin, M., boyd, d., Cody, R., and Herr-Stephenson, B., et al. (2009). *Hanging out, messing around, geeking out: Kids living and learning with new media.* Cambridge, MA: MIT Press.

ITU. (2008). Experts issue energy saving challenge. ITU-T Newslog. Retrieved February 18, 2009, from <http://www.itu.int/ITU-T/newslog/Experts+Issue+Energy+Saving+Challenge.aspx>.

Jenkins, H. (2006). Confronting the challenges of participatory culture: Media education for the 21st century. MacArthur Foundation. Retrieved August 1, 2009, from <http://www.digitallearn ing.macfound.org/atf/cf/%7BE45C7E0-A3E0-4B89-AC9C-E807E1B0AE4E%7D/JENKINS_ WHITE_PAPER.PDF>.

Jensen, R. (2007). The digital provide: Information (technology), market performance, and welfare in the south Indian fisheries sector. *Quarterly Journal of Economics, 122*(3), 879–924.

Kabel, M. (2006, July 18). Wal-Mart, critics slam each other on Web. *Washington Post.* Retrieved March 19, 2009, from <http://www.washingtonpost.com/wp-dyn/content/article/2006/07/18/ AR2006071800981.html>.

Kaboose.com. (2008). Earth Day games: Play free online games for moms and kids. Kaboose.com. Retrieved May 5, 2008, from <http://resources.kaboose.com/games/earthday.html>.

Kaiper, G. (2004). U.S. energy flow trends—2002 (UCRL-TR-129990-02). Retrieved August 1, 2009, from <https://publicaffairs.llnl.gov/news/energy/content/energy/energy_archive/energy_ flow_2002/USEnFlow02-quads.pdf>.

Keith, D. W., DeCarolis, J. F., Denkenberger, D. C., Lenschow, D. H., Malyshev, S. L., Pacala, S., et al. (2004). The influence of large-scale wind power on global climate. *Proceedings of the National Academy of Sciences of the United States of America, 101*(46), 16115–16120.

Keller, J. (2009). Energy drain by computers stifles efforts at cost control. Chronicle.com. Retrieved February 17, 2009, from <http://chronicle.com/free/v55/i18/18a00103.htm>.

Kincaid, J. (2009). Revealed: How the times got confused about Google and the tea kettle. Retrieved February 16, 2009, from <http://www.techcrunch.com/2009/01/12/revealed-the-times-made-up-that-stuff-about-google-and-the-tea-kettles/>.

Kinkade, S., & Verclas, K. (2008). *Wireless technology for social change: Trends in mobile use by NGOs.* Washington, DC: UN Foundation–Vodafone Group Foundation Partnership. Retrieved August 1, 2009, from <http://mobileactive.org/files/MobilizingSocialChange_full.pdf>.

Kirby, A. (2001, May 21). Annan slams Bush on global warming. BBC. Retrieved May 5, 2008, from <http://news.bbc.co.uk/1/hi/sci/tech/1341421.stm>.

Kleimaier, A., & Schröder, D. (2004). Hybrid cars, optimization, and control. In *Industrial technology, 2004. IEEE ICIT '04. 2004 IEEE international conference* (Vol. 2, pp. 1084–1089).

Klein, D. (1968). The introduction, increase, and crash of reindeer in St. Matthew Island. *Journal of Wildlife Management, 32*(2), 350–367.

Kloeppel, J. (2006). Weight gain of U.S. drivers has increased nation's fuel consumption. Retrieved February 17, 2009, from <http://news.illinois.edu/NEWS/06/1024auto.html>.

Klopfer, E., Squire, K., & Jenkins, H. (2002). Environmental detectives: PDAs as a window into a virtual simulated world. In *Proceedings IEEE international workshop on wireless and mobile technologies in education* (pp. 95–98). IEEE Computer Society. Retrieved May 5, 2008, from <http://portal.acm.org/citation.cfm?id=674392>.

Kranzberg, M. (1986). Technology and history: "Kranzberg's laws. *Technology and Culture, 27*(3), 544–560.

Kuhn, T. S. (1996). *The structure of scientific revolutions.* Chicago: University of Chicago Press.

Lakoff, G., & Johnson, M. (1980). *Metaphors we live by.* Chicago: University of Chicago Press.

Lambertson, L. (2009). Rain drain: Choosing a surf site after a storm. Retrieved February 16, 2009, from <http://www.exploratorium.edu/theworld/surfing/raindrain.html>.

Lampe, C., Johnston, E., & Resnick, P. (2007). Follow the reader: Filtering comments on slashdot. In *Proceedings of the SIGCHI conference on human factors in computing systems* (pp. 1253–1262). San Jose, CA: ACM.

lcv.org. (2009). Policy and Action Center. Retrieved February 17, 2009, from <http://action.lcv.org/lcv/policy_action.html>.

Leake, J., & Woods, R. (2009). Revealed: The environmental impact of Google searches. Times Online. Retrieved February 16, 2009, from <http://technology.timesonline.co.uk/tol/news/tech_and_web/article5489134.ece>.

Leakey, R. E., & Lewin, R. (1996). *The sixth extinction: Patterns of life and the future of humankind.* New York: Anchor Books.

Lechner, R. (2009). Greening of the Internet economy—Conference: Greening the power hungry data center. Retrieved February 15, 2009, from <http://greeninternet.calit2.net/>.

Leopold, A. (1966). *A Sand County almanac.* Oxford: Oxford University Press.

Lerner, J. (2009, January). Baby, you can share my car. *Delta Sky.* Retrieved February 16, 2009, from <http://www.delta-sky.com/sections/index.php/lifestyle/baby_you_can_share_my_car/>.

Leshed, G., Velden, T., Rieger, O., Kot, B., & Sengers, P. (2008). In-car GPS navigation: Engagement with and disengagement from the environment. In Burnett, M., et al. (Eds.), *Proceeding of the twenty-sixth annual SIGCHI conference on human factors in computing systems* (pp. 1675–1684). Florence: ACM.

Li, Y., & Landay, J. A. (2008). Activity-based prototyping of ubicomp applications for long-lived, everyday human activities. In *Proceeding of the twenty-sixth annual SIGCHI conference on human factors in computing systems* (pp. 1303–1312). Florence, Italy: ACM.

Libutti, P. O. (1997). Model organizational structures and best practices for successful national collaborative information partnerships. Paper presented at the USEIN Kick-Off Conference, Washington, DC.

Lovelock, J. (2006). *The revenge of Gaia*. New York: Basic Books.

Lovins, A. (2003). Some missing elements of sustainable development. Digital Vision Fellows, Stanford University. Retrieved August 1, 2009, from <http://www.rmi.org/images/PDFs/Security/S03-12_MissElemSusDev.pdf>.

Luo, J. (1995). Students rally for Mitsubishi boycott. *Daily Bruin*. Retrieved February 17, 2009, from <http://dailybruin.com/archives/id/4929/>.

Machiavelli, N. (1532). *The prince* (chapter VI). Retrieved February 16, 2009, from <http://www.constitution.org/mac/prince06.htm>.

Mankoff, J., Matthews, D., Fussell, S. R., & Johnson, M. (2007). Leveraging social networks to motivate individuals to reduce their ecological footprints. In *Proceedings of the 40th annual Hawaii international conference on system sciences* (p. 87). IEEE Computer Society. Retrieved May 8, 2008, from <http://portal.acm.org/citation.cfm?id=1255497>.

Markoff, J. (2006, September 26). Google to push for more electrical efficiency in PC's. *New York Times*. Retrieved May 8, 2008, from <http://www.nytimes.com/2006/09/26/technology/26google.html>.

Martin, A. (2006, November 6). The package may say healthy, but this grocer begs to differ. *New York Times*. Retrieved August 1, 2009, from <http://www.nytimes.com/2006/11/06/business/06grocery.html>

Mazmanian, D. (2009). *Toward sustainable communities: Transition and transformations in environmental policy* (2nd ed.). Cambridge, MA: MIT Press.

McDonough, W., & Braungart, M. (2002). *Cradle to cradle: Remaking the way we make things*. New York: North Point Press.

McPeak, S. (2009). Greening of the Internet economy—Conference: California's AB 32 and information communication technology. Retrieved February 15, 2009, from <http://greeninternet.calit2.net/>.

Meacham, J. (2009). Greening of the Internet economy—Conference: ICT and smart buildings. Retrieved February 15, 2009, from <http://greeninternet.calit2.net/>.

Meier, S. (2005). Civilization IV. Retrieved February 20, 2009, from <http://www.2kgames.com/civ4/home.htm>.

Millennium Ecosystem Assessment Board. (2005). *Living beyond our means: Natural assets and human well-being*. Retrieved August 1, 2009, from <http://www.millenniumassessment.org/documents/document.429.aspx.pdf>.

Mingay, S. (2007). Green IT: A new industry shock wave. Retrieved August 1, 2009, from <http://www.ictliteracy.info/rf.pdf/Gartner_on_Green_IT.pdf>.

Moher, T., Hussain, S., Halter, T., & Kilb, D. (2005). RoomQuake: Embedding dynamic phenomena within the physical space of an elementary school classroom. In *CHI '05 extended abstracts on human factors in computing systems* (pp. 1665–1668). Portland, OR: ACM.

Moher, T., Uphoff, B., Bhatt, D., Silva, B. L., & Malcolm, P. (2008). WallCology: Designing interaction affordances for learner engagement in authentic science inquiry. In *Proceeding of the Twenty-Sixth Annual SIGCHI Conference on Human Factors in Computing Systems* (pp. 163–172). Florence: ACM.

montereybayaquarium.org. (2009). Seafood watch recommendations for your iPhone. Monterey Bay Aquarium. Retrieved March 2, 2009, from <http://www.montereybayaquarium.org/cr/Sea foodWatch/web/sfw_iPhone.aspx>.

NakedWiki.org. (2008). Previous rides data—World naked bike ride (WNBR)—Wiki information site. Retrieved May 4, 2008, from <http://nakedwiki.org/index.php?title=Previous_rides_data>.

Nardi, B., & O'Day, V. (1999, May 3). Information ecologies: Using technology with heart: Chapter four: Information ecologies. Retrieved May 4, 2008, from <http://www.firstmonday.org/issues/issue4_5/nardi_chapter4.html>.

Nathan, L. P., Klasnja, P. V., & Friedman, B. (2007). Value scenarios: A technique for envisioning systemic effects of new technologies. In *CHI '07 extended abstracts on human factors in computing systems* (pp. 2585–2590). San Jose, CA: ACM.

National Academy of Sciences. (1995). National science education standards. Retrieved May 5, 2008, from <http://www.nap.edu/readingroom/books/nses/6a.html>.

National Environmental Education Foundation. (2001). Summary of Roper report 2001: Lessons from the environment. Retrieved August 1, 2009, from <http://www.neefusa.org/resources/roper2001-a.htm>.

National Fuel Cell Research Center. (2009). NFCRC projects: ZEV-NET. Retrieved February 16, 2009, from <http://www.nfcrc.uci.edu/2/ACTIVITIES/PROJECTS/ZEVNET/>.

National Public Radio. (2001). Coltan mining and eastern Congo's gorillas. Retrieved February 16, 2009, from <http://www.npr.org/programs/re/archivesdate/2001/dec/20011220.coltan.html>.

National Public Radio. (2008). Tuna: A favorite fish faces dangerous depletion. Retrieved February 16, 2009, from <http://www.npr.org/templates/story/story.php?storyId=94299340>.

nativeenergy.com. (2008). Carbon offsets for people and planet. NativeEnergy. Retrieved February 17, 2009, from <http://www.nativeenergy.com/>.

Nature Conservancy. (2009). Carbon footprint calculator: What's my carbon footprint? Retrieved February 17, 2009, from <http://www.nature.org/initiatives/climatechange/calculator/>.

Negroponte, N. (1991). Beyond the desktop metaphor. In A. Meyer, J. Guttag, P. Rivest, P. Szolovits (Eds.), *Research directions in computer science: An MIT perspective* (pp. 183–190). Cambridge, MA: MIT Press.

Negroponte, N. (1996). *Being digital* (1st ed.). New York: Vintage Books.

Nielsen, J. (1993). Iterative user-interface design. *IEEE Computer, 26*(11), 32–41.

Nobel Foundation. (2008). Peace 2007. Retrieved May 4, 2008, from <http://nobelprize.org/nobel_prizes/peace/laureates/2007/>.

Nokia. (2009). Substance and material management—Our responsibility—Environment. Retrieved February 16, 2009, from <http://www.nokia.com/A41041092>.

Norman, D. (1988). *The psychology of everyday things*. New York: Basic Books.

NRDC.org. (2004, August 27). NRDC: Forest facts. Retrieved May 4, 2008, from <http://www.nrdc.org/land/forests/fforestf.asp>.

OECD. (2009). Extended producer responsibility. Retrieved February 16, 2009, from <http://www.oecd.org/document/19/0,3343,en_2649_34281_35158227_1_1_1_1,00.html>.

OED.com. (1989a). Education. *Oxford English Dictionary*. Retrieved May 5, 2008, from <http://dictionary.oed.com/cgi/entry/50072205?single=1&query_type=word&queryword=education&first=1&max_to_show=10>.

OED.com. (1989b). Technology. *Oxford English Dictionary*. Retrieved February 16, 2009, from <http://dictionary.oed.com/cgi/entry/50248096?single=1&query_type=word&queryword=technology&first=1&max_to_show=10>.

Oliver, R. (2008, March 17). All about: Food and fossil fuels. CNN.com. Retrieved May 5, 2008, from <http://edition.cnn.com/2008/WORLD/asiapcf/03/16/eco.food.miles/>.

Olson, D. (2005). Species extinction rate speeding up. Retrieved January 2, 2009, from <http://news.minnesota.publicradio.org/features/2005/01/31_olsond_biodiversity/>.

Olson, G., & Olson, J. (2003). Mitigating the effects of distance on collaborative intellectual work. *Economics of Innovation and New Technology, 12*(1), 27–42.

Orkustofnun. (2005). *Energy statistics in Iceland*. Retrieved February 15, 2009, from <http://www.os.is/Apps/WebObjects/Orkustofnun.woa/swdocument/3879/Orkut%C3%B6lur+enska+2005.pdf>.

Overby, P. (2005). Environmental groups launch Exxon boycott. National Public Radio. Retrieved February 17, 2009, from <http://www.npr.org/templates/story/story.php?storyId=4749052>.

Parliamentary Office of Science and Technology. (2002). *Access to energy in developing countries*. Retrieved February 15, 2009, from <http://www.parliament.uk/post/pn191.pdf>.

Patterson, D. (2007). LUCI blog: 1 videoconference. 37 people. 122.7 tons of C02 saved. Retrieved February 16, 2009, from <http://luci.ics.uci.edu/blog/archives/2007/12/1_videoconferen_1.html>.

PBS.org. (2008). Nature: Interactives. Public Broadcasting Service. Retrieved May 5, 2008, from <http://www.pbs.org/wnet/nature/fun.html>.

Petit, C., & Petit, S. (2008). Piggyback. Retrieved February 16, 2009, from <http://androidapplications.com/apps/detail/piggyback>.

Phillips, C. (2003). Environmental group urges boycott of Royal Caribbean Cruise Line. Transportation > marine transportation, from AllBusiness.com. Retrieved February 17, 2009, from <http://www.allbusiness.com/environment-natural-resources/environmental/10221396-1.html>.

Pimentel, D., Garnick, E., Berkowitz, A., Jacobson, S., Napolitano, S., Black, P., et al. (1980). Environmental quality and natural biota. *Bioscience, 30*(11), 750–755.

Pimentel, D., & Pimentel, M. (2003). Sustainability of meat-based and plant-based diets and the environment. *American Journal of Clinical Nutrition, 78*(3), 660S–663S.

pledgebank.com. (2008). Successful pledges, closed to new signers. PledgeBank. Retrieved February 17, 2009, from <http://www.pledgebank.com/list/succeeded_closed?offset=50&sort=category>.

Polling Report. (2008). Environment. Retrieved May 4, 2008, from <http://www.pollingreport.com/enviro.htm>.

Pratt, R., Carpenter, F. L., & Tomlinson, B. (2006). The EcoRaft project: An interdisciplinary approach to teaching lessons in ecological restoration. *Bulletin of the Ecological Society of America, 87*(2), 176–182.

pricegrabber.com. (2009). ShopGreen. Retrieved March 3, 2009, from <http://shopgreen.pricegrabber.com/>.

P3 International Corporation. (2008). Kill a watt. P3. Retrieved February 17, 2009, from <http://www.p3international.com/products/special/P4400/P4400-CE.html>.

Quinn, D. (1997). *The story of B*. New York: Bantam Books.

Quran 31:20. (n.d.). Retrieved March 19, 2009, from <http://www.cs.cmu.edu/afs/cs.cmu.edu/project/cmt/resources/Quran/031.Luqman>.

Radon Labs. (2008). Radon Labs game development. Retrieved February 17, 2009, from <http://www.radonlabs.de/gamesgenius-biology.html>.

Ramsey, D. (2008). UC San Diego's GreenLight project to improve energy efficiency of computing. Retrieved February 17, 2009, from <http://ucsdnews.ucsd.edu/newsrel/science/07-08GreenLightProj.asp>.

raptureready.com. (2008). The rapture index. Retrieved May 5, 2008, from <http://www.raptureready.com/rap2.html>.

ratemyprofessors.com. (2009). Rate My Professors. Retrieved March 3, 2009, from <http://www.ratemyprofessors.com>.

Read, D. (2006). Working memory: A cognitive limit to non-human primate recursive thinking? *Cognitive Science Journal Archive*, 2674–2679.

Recker, W. (2009). Greening of the Internet economy—Conference: ICT and intelligent transportation. Retrieved February 15, 2009, from <http://greeninternet.calit2.net/>.

Regnier, E., & Tovey, C. (2007). Time horizons of environmental versus non-environmental costs: Evidence from US tort lawsuits. *Business Strategy and the Environment, 16*(4), 249–265.

Reuters. (2007, November 29). Global cellphone penetration reaches 50 pct: Technology, media, & telecommunications. Reuters.co.uk. Retrieved May 8, 2008, from <http://investing.reuters.co.uk/news/articleinvesting.aspx?type=media&storyID=nL29172095>.

Rocky Mountain Institute. (2009). Mission statement. Retrieved February 16, 2009, from <http://www.rmi.org/sitepages/pid55.php>.

Roberts, G. (1998). Competitive altruism: From reciprocity to the handicap principle. *Proceedings of the Royal Society B: Biological Sciences, 265*(1394), 427–431.

Rooney, B. (2007). ABC News: UPS figures out the "right way" to save money, time, and gas. Retrieved February 16, 2009, from <http://abcnews.go.com/wnt/story?id=3005890>.

Rosenbloom, S., & Barbaro, M. (2009, January 24). Green-light specials, now at Wal-Mart. *New York Times*. Retrieved February 16, 2009, from <http://www.nytimes.com/2009/01/25/business/25walmart.html?pagewanted=2&_r=1&hp>.

Rosenthal, E. (2008, November 27). Investment funds push an environmental agenda. *New York Times*. Retrieved February 16, 2009, from <http://www.nytimes.com/2008/11/28/business/28green.html>.

saisriskandarajah.com. (2009). Happy meat. Retrieved February 17, 2009, from <http://saisriskandarajah.com/happymeat/>.

Schults, C. (2006). GreenScanner. Grist.org. Retrieved March 3, 2009, from <http://gristmill.grist.org/story/2006/4/19/10053/3031>.

Secretariat of the Convention on Biological Diversity. (2006). *Global biodiversity outlook 2*. Retrieved May 4, 2008, from <http://www.cbd.int/doc/gbo2/cbd-gbo2-en.pdf>.

Serageldin, I. (2006). State of the planet 2006. Retrieved February 15, 2009, from <http://www.earth.columbia.edu/sop2006/transcripts/tr_serageldin_i.html>.

Shirky, C. (2008). Gin, television, and social surplus. Here Comes Everybody. Retrieved February 16, 2009, from <http://www.herecomeseverybody.org/2008/04/looking-for-the-mouse.html>.

Sightline Institute. (2009). The Cascadia scorecard. Retrieved February 16, 2009, from <http://scorecard.sightline.org/>.

Simon, H. A. (1996). *The sciences of the artificial*. Cambridge, MA: MIT Press.

SixNations.org. (2007, January 10). Haudenosaunee. Retrieved May 8, 2008, from <http://web.archive.org/web/20070110081428/http://www.sixnations.org/Culture/?article=seventh_generation>.

slashdot.org. (2008). GPS used to find graves in eco-burial sites. Slashdot. Retrieved February 17, 2009, from <http://news.slashdot.org/article.pl?sid=08/04/22/2148252>.

Smarr, L. L., Chien, A. A., DeFanti, T., Leigh, J., & Papadopoulos, P. M. (2003). The OptIPuter. *Communications of the ACM, 46*(11), 58–67.

Smart Growth America. (2009). Smart growth America: Growing cooler. Retrieved February 16, 2009, from <http://smartgrowthamerica.org/gcindex.html>.

Smith, D., Kay, A., Raab, A., & Reed, D. (2003). Croquet: A collaboration system architecture. Retrieved August 1, 2009, from <www.opencroquet.org/images/2/2b/2003_Croquet_Collab_Arch .pdf>.

snopes.com. (2007). The barometer problem. Retrieved February 16, 2009, from <http://www .snopes.com/college/exam/barometer.asp>.

Sorensen, V. (2002). Art and science: Challenges and possibilities. Federal University of Rio de Janeiro (UFRJ), Brazil. Retrieved February 16, 2009, from <http://visualmusic.org/text/VS-UFRJ .htm>.

Speth, J. G. (2008). *The bridge at the edge of the world: Capitalism, the environment, and crossing from crisis to sustainability.* New Haven, CT: Yale University Press.

Stephens-Romero, S., & Samuelsen, G. S. (2009). Demonstration of a novel assessment methodology for hydrogen infrastructure deployment. *International Journal of Hydrogen Energy, 34*(2), 628–641.

Sterman, J. D. (2008). ECONOMICS: Risk communication on climate: Mental models and mass balance. *Science, 322*(5901), 532–533.

StopGlobalWarming.org. (2009). StopGlobalWarming.org. Retrieved February 17, 2009, from <http://www.stopglobalwarming.org/>.

Strand, G. (2008). Keyword: Evil. *Harper's Magazine.* Retrieved February 16, 2009, from <http:// harpers.org/media/slideshow/annot/2008-03/index.html>.

Suchman, L. (1987). *Plans and situated actions: The problem of human-machine communication.* Cambridge: Cambridge University Press.

Sullivan, B. (1999). WTO protest spills onto the Web. Tech News on ZDNet. Retrieved February 17, 2009, from <http://news.zdnet.com/2100-9595_22-104379.html>.

Sun. (2008, November 29). China earmarks $73 mln for rural environment protection. *China View.* Retrieved February 16, 2009, from <http://news.xinhuanet.com/english/2008-11/29/ content_10428455.htm>.

Surfrider.org. (2009). Maine—Ocean illness database. Retrieved February 16, 2009, from <http:// www.surfrider.org/nne/index.php?page=76>.

Szewczyk, R., Osterweil, E., Polastre, J., Hamilton, M., Mainwaring, A., & Estrin, D. (2004). Habitat monitoring with sensor networks. *Communications of the ACM, 47*(6), 34–40.

Taylor, W. R. J., Terlouw, D. J., Olliaro, P. L., White, N. J., Brasseur, P., & ter Kuile, F. O. (2006). Use of weight-for-age-data to optimize tablet strength and dosing regimens for a new fixed-dose artesunate-amodiaquine combination for treating falciparum malaria. *Bulletin of the World Health Organization, 84*(12), 956–964.

Thackara, J. (2005). *In the bubble*. Cambridge, MA: MIT Press.

Thompson, C. (2005). Meet the life hackers. *New York Times*. Retrieved February 17, 2009, from <http://www.nytimes.com/2005/10/16/magazine/16guru.html>.

Thompson, M., & Moher, T. (2006). HelioRoom: Problem-solving in a whole class visual simulation. In *Proceedings of the 7th International Conference on Learning Sciences* (pp. 1000–1001). Bloomington, IN: International Society of the Learning Sciences. Retrieved May 5, 2008, from <http://portal.acm.org/citation.cfm?id=1150219>.

Tilkin, D., & KATU Web Staff (2008, November 5). How too much wind power may hurt salmon. Retrieved February 15, 2009, from <http://www.katu.com/news/33967994.html>.

timelinks.org. (2009). 60.jpg. Retrieved March 5, 2009, from <http://www.timelinks.org/uploads/images/gallery/60.jpg>.

Tomlinson, B. (2008a). Prototyping a community-generated, mobile device-enabled database of environmental impact reviews of consumer products. In *Proceedings of the 41st Annual Hawaii International Conference on System Sciences* (pp. 154ff). Washington, DC: IEEE.

Tomlinson, B. (2008b). Sustainably ours: A call for pro-environmental conspicuous consumption in the online world. *Interactions, 15*(6), 42–45.

Tomlinson, B. (2009). Broadening human horizons through green IT. In *ACM CHI 2009 Conference on Human Factors in Computing Systems, Workshop on Defining the Role of HCI in the Challenges of Sustainability*. Retrieved August 1, 2009, from <http://elainehuang.com/CHI-2009/p19-tomlinson.pdf>.

Tomlinson, B., Baumer, E., & Yau, M. L. (2006). The island metaphor. Paper presented at International Conference on Computer Graphics and Interactive Techniques (ACM SIGGRAPH 2006 Research posters), Boston, MA.

Tomlinson, B., Baumer, E., Yau, M. L., Carpenter, F. L., & Black, R. (2008). A participatory simulation for informal education in restoration ecology. *E-Learning, 5*(3), 238–255.

Tomlinson, B., Yau, M. L., & Baumer, E. (2006). Embodied mobile agents. In *Proceedings of the Fifth International Joint Conference On Autonomous Agents and Multiagent Systems* (pp. 969–976). Hakodate, Japan: ACM.

Tomlinson, B., Yau, M. L., Baumer, E., Goetz, S., Carpenter, L., Pratt, R., et al. (2006). The EcoRaft project: A multi-device interactive graphical exhibit for learning about restoration ecology. In *CHI '06 extended abstracts on human factors in computing systems* (pp. 1445–1450). Montreal: ACM.

Tomlinson, B., Yau, M. L., Baumer, E., Ross, J., Correa, A., & Ji, G. (2009). Richly connected systems and multi-device worlds. *Presence, 18*(1), 54–71.

treehugger.com. (2009). TreeHugger. Retrieved February 17, 2009, from <http://www.treehugger.com/>.

treehuggertv. (2009, March 5). Saul Griffith talks about climate change and renewable energy at greener gadgets (pt 2). Retrieved March 8, 2009, from <http://www.youtube.com/watch?v=rbj5k OCiiNk>.

Trope, Y., Liberman, N., & Wakslak, C. (2007). Construal levels and psychological distance: Effects on representation, prediction, evaluation, and behavior. *Journal of Consumer Psychology*, *17*(2), 83–95.

UN Economic and Social Commission for Asia and the Pacific. (2008). *Information and communications technology for food security and sustainable agriculture in the knowledge economy*. Retrieved February 16, 2009, from <http://www.unescap.org/icstd/cict/2008/CICT_INF1E.pdf>.

UN Environment Programme. (2005). Millennium ecosystem assessment. Retrieved May 5, 2008, from <http://www.millenniumassessment.org/en/About.aspx>.

UNESCO. (2002, January 6). A dynamic balance. Retrieved February 16, 2009, from <http://portal.unesco.org/en/ev.php-URL_ID=4029&URL_DO=DO_TOPIC&URL_SECTION=201.html>.

UNESCO. (2005). Education: Welcome to education for sustainable development. UNESCO. Retrieved May 5, 2008, from <http://portal.unesco.org/education/en/ev.php-URL_ID=27234&URL_DO=DO_TOPIC&URL_SECTION=201.html>.

UNESCO. (2007). United Nations Educational, Scientific, and Cultural Organization. UNESCO. org. Retrieved May 5, 2008, from <http://portal.unesco.org/education/en/ev.php-URL_ID=48781&URL_DO=DO_TOPIC&URL_SECTION=201.html>.

UNESCO & UNEP. (2002). *Cultural diversity and biodiversity for sustainable development*. Retrieved February 15, 2009, from <http://unesdoc.unesco.org/images/0013/001322/132262e.pdf>.

UN General Assembly. (1948, December 10). Universal declaration of human rights. Retrieved May 5, 2008, from <http://www.un.org/Overview/rights.html>.

UN General Assembly. (2000, September 18). Millennium declaration. Retrieved May 5, 2008, from <http://www.un.org/millennium/declaration/ares552e.pdf>.

United Nations. (2009). United Nations millennium development goals. Retrieved March 2, 2009, from <http://www.un.org/millenniumgoals/>.

United Nations. (2004). *World population to 2300*. Retrieved February 15, 2009, from <http://un.org/esa/population/publications/longrange2/WorldPop2300final.pdf>.

UN Secretariat. (1999). The world at six billion. Retrieved February 14, 2009, from <http://www.un.org/esa/population/publications/sixbillion/sixbillion.htm>.

upcdatabase.com. (2009). UPC database. Retrieved March 3, 2009, from <http://www.upcdatabase.com/>.

urbandictionary.com. (2009). Urban Dictionary. Retrieved March 3, 2009, from <http://www.urbandictionary.com/>.

U.S. Bureau of Labor Statistics. (2006, November 14). Consumer expenditures in 2004. Retrieved August 1, 2009 from <http://www.bls.gov/cex/csxann04.pdf>.

U.S. Census. (1998). Population of the 100 largest cities and other urban places in the United States: 1790 to 1990. Retrieved February 15, 2009, from <http://www.census.gov/population/www/documentation/twps0027/twps0027.html>.

U.S. Census. (2009a). Population estimates. Retrieved August 1, 2009, from <http://www.census.gov/popest/national/asrh/NC-EST2007/NC-EST2007-02.xls>.

U.S. Census. (2009b). U.S. and world population clocks—POPClocks. Retrieved February 15, 2009, from <http://www.census.gov/main/www/popclock.html>.

U.S. Congress (2009). H.R. 1—111th Congress: American Recovery and Reinvestment Act of 2009. Retrieved August 1, 2009, from <http://www.govtrack.us/congress/bill.xpd?bill=h111-1>.

U.S. Department of Energy. (2006). *World consumption of primary energy by energy type and selected country groups, 1980–2004*. Retrieved August 1, 2009, from <http://www.eia.doe.gov/pub/international/iealf/table18.xls>.

U.S. Environmental Protection Agency. (2006, November 15). Environmentally preferable purchasing. Retrieved June 22, 2008, from <http://www.epa.gov/opptintr/epp/pubs/about/about.htm>.

U.S. Environmental Protection Agency. (2007). *Municipal Solid Waste Generation, Recycling, and Disposal in the United States: Facts and Figures for 2006*. Retrieved August 7, 2009, from <http://www.epa.gov/osw/nonhaz/municipal/pubs/msw06.pdf>.

U.S. Environmental Protection Agency. (2008a). *Municipal solid waste in the United States: 2007 facts and figures*. Retrieved August 7, 2009, from <http://www.epa.gov/osw/nonhaz/municipal/pubs/msw07-rpt.pdf>.

U.S. Environmental Protection Agency. (2008b). EPA environmental kids club—Game room. Retrieved May 5, 2008, from <http://www.epa.gov/kids/game.htm>.

USGBC. (2008). USGBC: LEED. Retrieved February 21, 2009, from <http://www.usgbc.org/DisplayPage.aspx?CategoryID=19>.

U.S. Government Accountability Office. (2007). *Crude oil: Uncertainty about future oil supply makes it important to develop a strategy for addressing a peak and decline in oil production*. Retrieved August 1, 2009, from <http://www.gao.gov/new.items/d07283.pdf>.

U.S. National Science Foundation. (2006a, November 7). Informal Science Education (ISE) nsf08547. Retrieved May 5, 2008, from <http://www.nsf.gov/pubs/2008/nsf08547/nsf08547.htm>.

U.S. National Science Foundation. (2006b, November 7). Information and intelligent systems: Advancing human-centered computing, information integration and informatics, and robust intelligence nsf06572. Retrieved May 4, 2008, from <http://www.nsf.gov/pubs/2006/nsf06572/nsf06572.htm>.

Veblen, T. (1912). *The theory of the leisure class: An economic study of institutions*. London: Macmillan.

von Hippel, F. (2008, April). Nuclear fuel recycling: More trouble than it's worth. *Scientific American*. Retrieved February 27, 2009, from <http://www.sciam.com/article.cfm?id=rethinking -nuclear-fuel-recycling>.

von Weizsäcker, E. (2008, January 15). The climate challenge: Answers from technology, business, and society. Lecture at University of California, Irvine.

von Weizsäcker, E. U., Lovins, A. B., & Lovins, L. H. (1998). *Factor four: Doubling wealth—Halving resource use*. London: Earthscan.

Walker, B., Holling, C. S., Carpenter, S. R., & Kinzig, A. 2004. Resilience, adaptability and transformability in social–ecological systems. *Ecology and Society, 9*(2): 5. Retrieved August 10, 2009 from: <http://www.ecologyandsociety.org/vol9/iss2/art5>.

Watson, R. (2009, February 5). *GreenerBuildings News*. Retrieved March 29, 2009, from <http://www.greenerbuildings.com/newsletter/greenerbuildings/2009/02/05/newsletter>.

Weihl, B. (2009). Greening of the Internet economy—Conference: Keynote: A vision for a green IT future. Retrieved February 15, 2009, from <http://greeninternet.calit2.net/>.

wfp.org. (2008). Tsunami crisis: Rebuilding. WFP—Latest news—Humanitarian crisis. Retrieved May 5, 2008, from <http://www.wfp.org/english/?ModuleID=137&Key=1858>.

Wheeland, M. (2008). The top green computing stories of 2008. Kiplinger.com. Retrieved February 17, 2009, from <http://www.kiplinger.com/businessresource/summary/archive/2008/green tips_computing.html>.

Wilson, E. O. (1992). *The diversity of life*. Cambridge, MA: Belknap Press of Harvard University.

Winston, A. (2008). How smart companies use environmental strategy to innovate, create value, and build competitive advantage. Lecture presented at Innovation & Globalization in Green conference, October 1, 2008. Irvine, CA.

Woodruff, A., Hasbrouck, J., & Augustin, S. (2008). A bright green perspective on sustainable choices. In *Proceeding of the Twenty-sixth Annual SIGCHI Conference on Human Factors in Computing Systems* (pp. 313–322). New York: ACM.

World Bank. (2001). DEPweb: Life expectancy text 1. Retrieved February 16, 2009, from <http://www.worldbank.org/depweb/english/modules/social/life/index.html>.

WorldNakedBikeRide.com. (2007). World naked bike ride (WNBR)—Naked bicycle people power! Stop indecent exposure to vehicle emissions! Retrieved May 4, 2008, from <http://worldnaked bikeride.org/index.html>.

Wright, S. (1922). Coefficients of inbreeding and relationship. *American Naturalist, 56*(645), 330–338.

Yeh, R., Liao, C., Klemmer, S., Guimbretière, F., Lee, B., Kakaradov, B., et al. (2006). ButterflyNet: A mobile capture and access system for field biology research. In *Proceedings of the SIGCHI conference on human factors in computing systems* (pp. 571–580). Montreal: ACM.

Young, J. (2009). Greening of the Internet economy—Conference: ICT and smart buildings: Building management systems, communications platforms, and more. Retrieved February 15, 2009, from <http://greeninternet.calit2.net/>.

Youngman, S. (2008, November 18). Obama talks climate change with governors. *Hill*. Retrieved February 16, 2009, from <http://thehill.com/leading-the-news/obama-talks-climate-change-with-governors-2008-11-18.html>.

Zahavi, A., & Zahavi, A. (1999). *The handicap principle: A missing piece of Darwin's puzzle*. Oxford: Oxford University Press.

Index